Disclaimer

The publisher of this book is by no way associated with the National Institute of Standards and Technology (NIST). The NIST did not publish this book. It was published by 50 page publications under the public domain license.

50 Page Publications.

Book Title: Documentation and Assessment of the Transport Property Models Implemented in NIST REFPROP (Version 8.0)

Book Author: Justin C. Chichester; Marcia L. Huber

Book Abstract: In this report, we describe an extended corresponding states model for viscosity and thermal conductivity of mixtures implemented in version 8 of the NIST computer program REFPROP. The model is a modification of a one-fluid, extended corresponding states (ECS) model for thermal conductivity and viscosity originally developed by Ely and Hanley (Ind. Eng. Chem. Fundam., 1981, 20:323-332). We apply the model to selected mixtures representative of the fluids available in the NIST 23 database REFPROP (v8) such as refrigerants and natural gas constituent fluids, and present comparisons with experimental data. Comparisons are given for both gas, liquid, and supercritical conditions.

Citation: NIST Interagency/Internal Report (NISTIR) - 6650

Keyword: extended corresponding states; natural gas; refrigerants; thermal conductivity; viscosity

NISTIR 6650

Documentation and Assessment of the Transport Property Model for Mixtures Implemented in NIST REFPROP (Version 8.0)

Justin C. Chichester
Marcia L. Huber

National Institute of Standards and Technology
U.S. Department of Commerce

NISTIR 6650

Documentation and Assessment of the Transport Property Model for Mixtures Implemented in NIST REFPROP (Version 8.0)

Justin C. Chichester
Marcia L. Huber

*Physical and Chemical Properties Division
Chemical Science and Technology Laboratory
National Institute of Standards and Technology
Boulder, CO 80305-3328*

June 2008

U.S. DEPARTMENT OF COMMERCE
Carlos M. Gutierrez, Secretary
NATIONAL INSTITUTE OF STANDARDS AND TECHNOLOGY
James M. Turner, Deputy Director

Contents

1. Introduction 2
2. Viscosity Model 3
   2.1 Pure Fluids 3
   2.2 Mixtures 6
3. Thermal Conductivity Model 8
   3.1 Pure Fluids 8
   3.2 Mixtures 11
4. Results 14
5. Conclusions 29
6. References 44

Documentation and Assessment of the Transport Property Model for Mixtures Implemented in NIST REFPROP (Version 8.0)

Justin Chichester and Marcia L. Huber
National Institute of Standards and Technology *
Boulder CO USA 80305-3328

In this report, we describe an extended corresponding states model for viscosity and thermal conductivity of mixtures implemented in Version 8 of the NIST computer program REFPROP. The model is a modification of a one-fluid, extended corresponding states (ECS) model for thermal conductivity and viscosity originally developed by Ely and Hanley (Ind. Eng. Chem. Fundam., 1981, 20:323-332). We apply the model to selected mixtures representative of the fluids available in the NIST 23 database REFPROP (v8), such as refrigerants and natural gas constituent fluids, and present comparisons with experimental data. Comparisons are given for gas, liquid, and supercritical conditions.

Keywords: extended corresponding states; natural gas; refrigerants; thermal conductivity; viscosity.

---

* Physical and Chemical Properties Division, Chemical Science and Technology Laboratory.

1. Introduction

The NIST Reference Fluid Thermodynamic and Transport Properties Database, NIST23 (REFPROP) [1], provides thermophysical properties for 84 pure fluids and for mixtures of up to 20 components. For pure fluids, it contains equation of state formulations for the thermodynamic properties, and models or correlations for the viscosity and thermal conductivity. The references for the individual pure fluid formulations are provided in the program itself through the "fluid information" screen. For mixtures, there is presently only one model available for the thermal conductivity and viscosity. It is based upon the extended corresponding states concept. In 1981, Ely and Hanley published a one-fluid, extended corresponding states model for viscosity [2], and in 1983 applied the model to thermal conductivity [3]. Since that time, there have been several modifications to the Ely-Hanley method, each improving upon the original model [4-10]. Variations of this model have been successfully implemented in two older NIST databases, NIST4 (Supertrapp) [11] and NIST14 (DDMIX) [12], and most recently in NIST23 (REFPROP) [13]. In this report we will describe the formulation used in the most recent version of the REFPROP computer program [1] and give comparisons with experimental data for the viscosity and thermal conductivity of selected mixtures of interest, including refrigerant mixtures and natural gas mixtures.

2. Viscosity Model

2.1 Pure Fluids

In the extended corresponding states model used here, we represent the viscosity of a pure fluid as a sum of a dilute gas and a residual contribution, and apply a corresponding states principle to the residual contribution only [2],

$$\eta(T,\rho) = \eta^*(T) + \Delta\eta(T,\rho) = \eta^*(T) + \Delta\eta_0(T_0,\rho_0)F_\eta(T,\rho), \tag{1}$$

where the superscript * denotes a dilute gas value, and the subscript 0 denotes a reference fluid value. The viscosity of the reference fluid is evaluated at a conformal temperature and density T_0 and ρ_0 given by

$$T_0 = T/f \tag{2}$$

and

$$\rho_0 = \rho h. \tag{3}$$

The quantities f and h are called equivalent substance reducing ratios, and relate the reference fluid to the fluid of interest using a ratio of critical parameters (denoted by the subscript c) and functions of temperature and density known as shape functions θ and ϕ,

$$f = \frac{T_c}{T_{c0}}\theta \tag{4}$$

and

$$h = \frac{\rho_{c0}}{\rho_c}\phi. \tag{5}$$

The shape factors can be considered functions of both temperature and density. For small, nonpolar, almost-spherical molecules they are nearly unity and can be thought of as functions to compensate for deviations from a spherical shape. Different approaches have been taken to determine shape factors [4, 5, 8] usually depending on the amount and quality of p,V,T information available for the fluid; in this work, we generally have available accurate formulations for the thermodynamic properties of the pure fluids, either in terms of a Helmholtz free energy equation or a pVT equation of state (EOS), and we use a form of the "exact" shape factor method [4]. It is a requirement in this method to first determine the thermodynamic shape factors.

The dilute-gas viscosity in Eq. (1) is given by the Chapman-Enskog theory [14]

$$\eta^*(T) = \frac{5\sqrt{mk_B T}}{16\pi\sigma^2 \Omega^{(2,2)}}, \tag{6}$$

where the dilute-gas viscosity is η^*, m is the molecular mass, k_B is the Boltzmann constant, and T is the absolute temperature. We will further assume that a Lennard-Jones 12-6 potential applies, and use the Lennard-Jones collision diameter for σ. Neufeld et al. [15] gave the following empirical correlation for the calculation of the collision integral $\Omega^{(2,2)}$

$$\Omega^{(2,2)} = 1.16145(T^*)^{-0.14874} + 0.52487 e^{-0.77320 T^*} + 2.16178 e^{-2.43787 T^*}, \tag{7}$$

with the dimensionless temperature $T^* = k_B T/\varepsilon$, and ε the minimum of the Lennard-Jones pair-potential energy. The range of validity of this empirical correlation is $0.3 < T^* < 100$.

The factor F_η in Eq. (1) is found by using the expression

$$F_\eta = f^{1/2} h^{-2/3} \left[\frac{M}{M_0}\right]^{1/2}, \qquad (8)$$

where M is the molar mass of the fluid and M_0 is the molar mass of the reference fluid. The model as developed to this point is predictive, and does not use any information on the viscosity of the fluid (except for the dilute-gas piece that requires Lennard-Jones ε and σ). The functions f and h are found from thermodynamic data and are described in Klein et al. [9]. In order to improve the representation of the viscosity, an empirical correction factor may be used if there are experimental viscosity data available. We then evaluate Eq. (1) at $\rho_{0,v}$ instead of ρ_0, where [9]

$$\rho_{0,v}(T,\rho) = \rho_0(T,\rho)\psi(\rho_r), \qquad (9)$$

and ψ is a polynomial in reduced density $\rho_r = \rho/\rho_c$ of the form

$$\psi(\rho_r) = \sum_{k=0}^{n} c_k \rho_r^k,$$

(10)

where the coefficients c_k are constants found from fitting the experimental viscosity data. As indicated in Eq. (1), in order to evaluate the viscosity of a particular fluid, the value of the residual viscosity of a reference fluid is required. It is not necessary to use the same reference fluid for all fluids. However, when the model is used in a predictive mode, it is best to select the reference fluid that is most similar in chemical nature to the fluid of interest. The reference fluid should also have a very accurate equation of state and viscosity surface. When using pure fluid experimental viscosity to essentially "correct" the viscosity, the choice of reference fluid is not as important, since an empirical correction factor determined from data is applied, as in Eqs. (9) through (10).

2.2 Mixtures

The extension of the model to mixtures is similar to that presented for pure fluids, but involves an extra step. First, one represents the properties of the mixture in terms of a hypothetical pure fluid (denoted here by the subscript x) that is obtained through the use of mixing rules,

$$f_x = h_x^{-1} \sum_{j=1}^{n} \sum_{i=1}^{n} x_i x_j f_{ij} h_{ij} \tag{11}$$

$$h_x = \sum_{j=1}^{n} \sum_{i=1}^{n} x_i x_j h_{ij}, \tag{12}$$

and combining rules,

$$f_{ij} = \sqrt{f_i f_j} (1 - k_{ij,f\eta}) \tag{13}$$

$$h_{ij} = \frac{\left(h_i^{1/3} + h_j^{1/3}\right)^3}{8} (1 - k_{ij,h\eta}), \tag{14}$$

where we introduce binary interaction parameters $k_{ij,f\eta}$ and $k_{ij,h\eta}$ that can be determined by fitting mixture experimental data when available. Otherwise, they are set to zero.

The iterative method used to determine f_i and h_i by use of equations of state is described in detail in the manuscript of Klein et al. [9]. In that work, the authors note that although the procedure for finding the shape factors is straightforward, it is complicated in practice and the iterative procedure can fail at both high densities and very low densities. In addition, we noted some failures with fuel mixtures in the supercritical region that occurred with the use of the Helmholtz equations of state, but did not occur with a much simpler equation of state such as the Peng-Robinson. We suspect that this might be due to in part to the iterations entering the two-phase region where the Helmholtz equations can have very complicated behavior. To partially alleviate this problem, we assume that at some high temperature, here selected as 1000 K, the individual pure-fluid shape factors will have a limiting value of

$$f_j = f_x \frac{T_{c,j}}{T_x} \qquad (15)$$

$$h_j = h_x \frac{\rho_x}{\rho_{c,j}}, \qquad (16)$$

where the subscript x denotes the mixture and j denotes the pure fluid. In the fluid region where the compressibility factor ($z=p/\rho RT$) of the fluid mixture is above 0.3, and the temperature is above the critical temperature of the mixture, we linearly interpolate the value of the f_j and h_j between its limiting value in Eqs. (15) through (16) and the value at the critical temperature.

We assume that the viscosity of the mixture obeys a corresponding states principle,

$$\eta(T,\rho,x) = \eta^*(T,x) + \Delta\eta(T,\rho,x) = \eta^*(T,x) + \Delta\eta_0(T_0,\rho_0)F_\eta(T,\rho), \qquad (17)$$

where the factor F_η for mixtures is found with

$$F_\eta = f_x^{1/2} h_x^{-2/3} g_x^{1/2}, \qquad (18)$$

$$g_x^{1/2} = \frac{\sum_{i=1}^{n}\sum_{j=1}^{n} x_i x_j f_{ij}^{1/2} h_{ij}^{4/3} M_{ij}^{1/2}}{f_x^{1/2} h_x^{4/3}} \qquad (19)$$

$$M_{ij} = \frac{2}{1/M_i + 1/M_j}. \qquad (20)$$

For the dilute gas viscosity, binary interaction parameters $k_{ij,\sigma}$ and $k_{ij,\varepsilon}$ may be used, where

$$\sigma_{ij} = (1 - k_{ij,\sigma})\sqrt{\sigma_i \sigma_j} \qquad (21)$$

$$\varepsilon_{ij}/k_B = (1 - k_{ij,\varepsilon})\sqrt{(\varepsilon_i/k_B)(\varepsilon_j/k_B)}. \qquad (22)$$

When there are sufficient dilute-gas viscosity data available, these parameters are obtained by fitting experimental data. Otherwise, they are set to zero.

3. Thermal Conductivity Model

3.1 Pure Fluids

We start with the procedure of Ely and Hanley [3] and represent the thermal conductivity of a fluid as the sum of translational (from collisions between molecules) and internal (due to internal motions of the molecule) modes of energy transfer,

$$\lambda(T,\rho) = \lambda^{int}(T) + \lambda^{trans}(T,\rho). \qquad (23)$$

The translational contribution may be further divided into a translational dilute-gas contribution (denoted here by a superscript *) that is a function only of temperature, a residual contribution, and a critical enhancement,

$$\lambda^{trans}(T,\rho) = \lambda^*(T) + \lambda^r(T,\rho) + \lambda^{crit}(T,\rho), \tag{24}$$

leading to the following expression for the thermal conductivity

$$\lambda(T,\rho) = \lambda^{int}(T) + \lambda^*(T) + \lambda^r(T,\rho) + \lambda^{crit}(T,\rho). \tag{25}$$

We use an Eucken correlation for the internal contribution

$$\lambda^{int}(T) = \frac{f_{int}\eta^*}{M}\left[C_p^* - \frac{5}{2}R\right], \tag{26}$$

where C_p^* is the ideal-gas heat capacity in J/(mol·K), R is the molar gas constant [16] (8.314 472 J/(mol·K)), η^* is the dilute-gas viscosity (µPa·s) as given in Eq. (6), f_{int} is set to 1.32·10^{-3}, and λ is in W/(m·K). If sufficient dilute-gas thermal conductivity data are available, f_{int} is fit to a polynomial in temperature,

$$f_{int} = a_0 + a_1 T. \tag{27}$$

For the dilute-gas translational contribution (in W/(m·K)) we use

$$\lambda^*(T) = \frac{15 \times 10^{-3} R\eta^*}{4M}, \tag{28}$$

where the dilute gas viscosity η^* is from Eq. (6). The residual contribution is found using extended corresponding states:

$$\lambda^r(T,\rho) = \lambda_0^r(T_0,\rho_0) F_\lambda, \tag{29}$$

with

$$F_\lambda = f^{1/2} h^{-2/3} \left[\frac{M_0}{M}\right]^{1/2}. \tag{30}$$

In order to improve the representation of the thermal conductivity, an empirical correction factor may be used if there are experimental thermal conductivity data available. We then evaluate Eq. (29) at $\rho_{0,k}$ instead of ρ_0, where [8]

$$\rho_{0,k}(T,\rho) = \rho_0(T,\rho)\chi(\rho_r), \tag{31}$$

and χ is a polynomial in reduced density $\rho_r = \rho/\rho_c$ of the form

$$\chi(\rho_r) = \sum_{k=0}^{n} b_k \rho_r^k, \tag{32}$$

where the coefficients b_k are found from fitting the experimental thermal conductivity data.

The critical contribution is computed using a simplified crossover model developed by Olchowy and Sengers [17],

$$\lambda^{\text{crit}}(T,\rho) = \frac{\rho C_p R_0 k_B T}{6\pi\eta\xi}(\Omega - \Omega_0), \tag{33}$$

where the heat capacity at constant pressure, $C_p(T,\rho)$, is obtained from the equation of state, $R_0 = 1.03$ is a universal constant [18], and the viscosity, $\eta(T,\rho)$, is obtained from the method described earlier. The crossover functions Ω and Ω_0 are determined by

$$\Omega = \frac{2}{\pi}\left[\left(\frac{C_p - C_v}{C_p}\right)\arctan(q_d\xi) + \frac{C_v}{C_p}(q_d\xi)\right], \tag{34}$$

$$\Omega_0 = \frac{2}{\pi}\left[1-\exp\left(\frac{-1}{(q_d\xi)^{-1}+\frac{1}{3}\left(\frac{(q_d\xi)\rho_c}{\rho}\right)^2}\right)\right]. \quad (35)$$

The heat capacity at constant volume, $C_v(T,\rho)$, is obtained from the equation of state, and the correlation length ξ is given by

$$\xi = \xi_0\left[\frac{p_c\rho}{\Gamma\rho_c^2}\right]^{\nu/\gamma}\left[\left.\frac{\partial\rho(T,\rho)}{\partial p}\right|_T - \frac{T_R}{T}\left.\frac{\partial\rho(T_R,\rho)}{\partial p}\right|_T\right]^{\nu/\gamma}. \quad (36)$$

The partial derivative of density with respect to pressure $\partial\rho/\partial p|_T$ is evaluated with the equation of state at the system temperature T and a reference temperature, T_R. The reference temperature is a value where the critical enhancement is assumed to be negligible: $T_R = 1.5T_c$. The exponents $\gamma = 1.239$ and $\nu = 0.63$ are universal constants [18]. The critical amplitudes Γ and ξ_0 are system-dependent and can be determined by the asymptotic behavior of the equation of state in the critical region. For most fluids, we set these to the fixed values $\Gamma = 0.0496$ and $\xi_0 = 1.94 \times 10^{-10}$ m. The thermal conductivity at the critical point itself is infinite. The cutoff wavenumber q_d (or alternatively, its inverse q_d^{-1}) is system-dependent and determined from regression of experimental data when available or estimated [19].

3.2 Mixtures

The extension of the model to mixtures is similar to that presented for viscosity,

$$\lambda(T,\rho,x) = \lambda^*(T,x) + \lambda^{int}(T,x) + \Delta\lambda^r(T,\rho) + \lambda^{crit}(T,\rho,x) \quad (37)$$

and only the residual contribution is treated with corresponding states.

$$\lambda(T,\rho,x) = \lambda^*(T,x) + \lambda^{int}(T,x) + \Delta\lambda_0(T_0,\rho_0)F_\lambda(T,\rho) + \lambda^{crit}(T,\rho,x). \quad (38)$$

The internal and translational dilute-gas contributions for the mixture are found with the empirical mixing rule,

$$\lambda_{\text{mix}}^{\text{int}}(T,x) + \lambda_{\text{mix}}^{*}(T,x) = \sum_{j=1}^{n} \frac{x_j (\lambda_j^{\text{int}}(T) + \lambda_j^{*}(T))}{\sum_{i=1}^{n} x_i \phi_{ji}} , \qquad (39)$$

with

$$\phi_{ji} = \frac{(1 - k_{ij,\lambda})\left(1 + \sqrt{\eta_j^{*}/\eta_i^{*}} (M_j/M_i)^{1/4}\right)^2}{(8(1 + M_j/M_i))^{1/2}} . \qquad (40)$$

All quantities are evaluated at the mixture temperature T, and the dilute-gas viscosity is found with Eq. (6). The parameter $k_{ij,\lambda}$ is an empirical binary interaction parameter for the dilute gas region determined by fitting experimental data.

The residual contribution for the mixture requires calculation of a mixture F_λ with

$$F_\lambda = f_x^{1/2} h_x^{-2/3} g_x^{1/2}, \qquad (41)$$

$$g_x^{1/2} = \frac{M_0^{1/2}}{f_x^{1/2} h_x^{4/3}} \sum_{i=1}^{n} \sum_{j=1}^{n} x_i x_j (f_{ij})^{1/2} \left(\frac{2}{1/g_i + 1/g_j}\right)^{-1/2} (h_{ij})^{4/3}, \qquad (42)$$

with g_i found with

$$g_i = M_0 \left(\frac{\lambda_0^{\text{r}}(T_0, \rho_0)}{\lambda_j^{\text{r}}(T_j, \rho_j)}\right)^2 f_j h_j^{-4/3} . \qquad (43)$$

Similar to the procedure for the calculation of the viscosity, we use the combining rules

$$f_{ij} = \sqrt{f_i f_j}(1-k_{ij,f\lambda}) \tag{44}$$

$$h_{ij} = \frac{\left(h_i^{1/3} + h_j^{1/3}\right)^3}{8}(1-k_{ij,h\lambda}), \tag{45}$$

where we introduce binary interaction parameters $k_{ij,f\lambda}$ and $k_{ij,h\lambda}$ that can be determined by fitting experimental data when available. Otherwise, they are set to zero.

As observed for pure fluids, there also is an enhancement of the thermal conductivity observed near the critical point; however, theory for mixtures is not as well developed as for pure fluids. In REFPROP 8 we make a simple approximation where we obtain the parameters necessary for the simplified Olchowy-Sengers model (Eqs. 33 through 36, above) by taking a mole fraction average,

$$T_R = 1.5\sum_{i=1}^{n} x_i T_{c_i} \tag{46}$$

$$q_d = \sum_{i=1}^{n} x_i q_{d_i} \tag{47}$$

$$\Gamma = \sum_{i=1}^{n} x_i \Gamma_i \tag{48}$$

$$\xi_0 = \sum_{i=1}^{n} x_i \xi_i \tag{49}$$

The other parameters γ, ν are universal and are unchanged from their pure-fluid values. The thermodynamic properties are evaluated at the mixture T and ρ, not at scaled values, and the mixture T_c, p_c and ρ_c are used. In addition, at the critical point the thermal conductivity of a pure fluid is infinite [20]; for a mixture, theory and experiment suggest that a finite value is achieved [21]. Therefore we have imposed a somewhat arbitrary upper limit (100 % of the background value) on the magnitude of the enhancement possible in a mixture.

4. Results

We tested the model on a limited set of mixture data containing both refrigerant mixtures and hydrocarbon mixtures. When possible, we obtained binary interaction parameters to improve the representation of the data. The parameters were obtained by least squares regression of experimental data. The resulting values for the binary interaction parameters for both the dilute and residual viscosity are given in Table 1 below. These binary interaction parameters are contained in the file HMX.BNC that is distributed with the REFPROP computer program.* In addition, we provide a list of the Lennard-Jones 12-6 parameters for pure fluids in REFPROP8 in Table 2. References for these values are provided, however, often it was necessary to estimate the values. The estimation methods of Chung et al. [22] and a corresponding-states estimation method [7] were used.

Table 3 gives the deviations of the model compared with experimental data for selected binary systems. We use the following definitions for average absolute deviation (AAD), bias, and root-mean-square (RMS) deviation:

$$AAD = \frac{100}{n} \sum_{i=1}^{n} \left| \frac{\eta_i^{cal}}{\eta_i^{exp}} - 1 \right|, \tag{50}$$

$$BIAS = \frac{100}{n} \sum_{i=1}^{n} \left(\frac{\eta_i^{cal}}{\eta_i^{exp}} - 1 \right), \tag{51}$$

and

$$RMS^2 = \frac{100}{n}\left(\sum_{i=1}^{n}\left(\frac{\eta_i^{cal}}{\eta_i^{exp}}-1\right)^2\right) - \left(\frac{100}{n}\sum_{i=1}^{n}\left(\frac{\eta_i^{cal}}{\eta_i^{exp}}-1\right)\right)^2. \qquad (52)$$

In general, the model represents the viscosity of the mixtures to within 5 % to 10 %. One notable exception is the data for the CO_2/decane system measured by Cullick and Mathis [23]. The model does not represent this system well even with the use of interaction parameters. This is probably due to the large size differences between the molecules. It has been noted previously [24] that one-fluid corresponding states types of models have difficulties representing systems where the molecules differ greatly in size. The trend of increasing deviations as the size difference between the molecules increases is also demonstrated by the data of Aucejo et al. [25] in the systems of pentane with increasingly larger alkanes, with the largest deviations observed for the pentane/dodecane system. Further work in this area is needed.

There also are large discrepancies between the model predictions and the viscosity data of Heide [26] for the binary mixture of R143a and R125. This system has also been studied by Laesecke [27] and is represented well by the model, so we suspect that there may have been an experimental problem with the Heide [26] data for this system.

Table 1a. Values of interaction parameters for viscosity for selected refrigerant, cryogen and hydrogen mixtures.

System	$k_{ij,\sigma}$	$k_{ij,\varepsilon}$	$k_{ij,f\eta}$	$k_{ij,h\eta}$	Source of data
R134a/R32	0.0437	-0.4678	-0.0645	-0.0305	[28] [29] [30]
R125/propane	-0.21	0.3776	0.0	0.0	[30]
R22/propane	-0.1284	0.3331	0.0	0.0	[31]
R115/propane	-0.2744	0.414	0.0	0.0	[31]
R32/propane	0.2857	-0.4346	0.0	0.0	[28]
R12/R22	-0.0213	0.0698	0.0	0.0	[31]
R13/R22	-0.0332	0.1389	0.0	0.0	[31]
R13B1/R22	-0.2505	0.7028	0.0	0.0	[31]
R14/R22	0.0523	-0.1194	0.0	0.0	[31]
R14/CO	-0.0421	-0.0638	0.0	0.0	[32]
R22/R152a	-0.1382	0.4355	0.0	0.0	[31]
R32/R124	0.0	0.0	-0.0813	-0.1158	[29]
R125/R134a	-0.0045	0.0038	-0.0829	-0.0106	[28] [29] [33]
hydrogen/methane	-0.2744	-0.1738	0.0	0.0	[22, 34]
hydrogen/nitrogen	-0.6225	0.1388	0.0	0.0	[34]
hydrogen/CO2	-0.9218	0.8531	0.0	0.0	[34]
hydrogen/oxygen	-0.7661	0.3683	0.0	0.0	[34]
hydrogen/CO	-0.7599	0.8536	0.0	0.0	[34]
hydrogen/helium	-0.0480	-1.9733	0.0	0.0	[34]
hydrogen/argon	-1.0538	0.8177	0.0	0.0	[34]
hydrogen/neon	-0.6105	-0.4449	0.0	0.0	[34]
hydrogen/ammonia	-0.5275	0.8419	0.0	0.0	[34]
hydrogen/ethylene	-0.7131	0.8778	0.0	0.0	[34]
hydrogen/propylene	-0.5744	0.7662	0.0	0.0	[34]
hydrogen/cis-butene	-0.7844	0.8518	0.0	0.0	[34]
helium/argon	-0.187	-2	0.0	0.0	[35]
krypton/CO	-0.08889	-0.05653	0.0	0.0	[32]

Table 1b. Values of interaction parameters for viscosity for selected natural gas component and hydrocarbon mixtures.

System	$k_{ij,\sigma}$	$k_{ij,\varepsilon}$	$k_{ij,f\eta}$	$k_{ij,h\eta}$	Source of data
methane/CO	-0.0165	-0.2493	0.0	0.0	[32]
methane/CO2	-0.00677	-0.3342	0.0	0.0	[36]
methane/ethane	0.0331	-0.2549	0.0	0.0	[37]
methane/propane	-0.0182	-0.0263	0.0	0.0	[37]
methane/butane	0.0030	-0.1292	0.0	0.0	[36]
methane/nitrogen	-0.3594	0.9076	-0.0612	0.0398	[38]
ethane/propane	0.0557	-0.2792	0.0	0.0	[37]
ethane/butane	0.0350	-0.1858	0.0	0.0	[37]
ethane/hydrogen	-0.4544	0.3877	0.0	0.0	[34]
propane/butane	0.0825	-0.3471	0.0	0.0	[37]
propane/hydrogen	-0.5926	0.7304	0.0	0.0	[34]
pentane/heptane	0.0	0.0	-0.0892	-0.1474	[25]
pentane/dodecane	0.0	0.0	-0.1188	-0.4911	[25]
pentane/decane	0.0	0.0	-0.0965	-0.2775	[25]
pentane/nonane	0.0	0.0	-0.1195	-0.1610	[25]
hexane/octane	0.0	0.0	-0.0519	-0.0010	[25]
hexane/dodecane	0.0	0.0	-0.0168	-0.2999	[25]
hexane/decane	0.0	0.0	-0.0550	-0.1310	[25]
hexane/nonane	0.0	0.0	-0.0263	-0.0738	[25]
hexane/cyclohexane	0.0	0.0	-0.12	0.2243	[39]
heptane/dodecane	0.0	0.0	-0.0716	-0.1600	[25]
heptane/decane	0.0	0.0	-0.0374	-0.0677	[25]
heptane/nonane	0.0	0.0	-0.0704	-0.0117	[25]
octane/dodecane	0.0	0.0	-0.0802	-0.1160	[25]
octane/decane	0.0	0.0	-0.0299	-0.0488	[25]
nonane/dodecane	0.0	0.0	-0.0634	-0.0803	[25]
nonane/decane	0.0	0.0	-0.0671	-0.0241	[25]
decane/dodecane	0.0	0.0	-0.0499	-0.0614	[25]

Table 2. Lennard-Jones 12-6 parameters for selected pure fluids in REFPROP v8.

Fluid	ε/k_B(K)	σ(nm)	Ref.	Fluid	ε/k_B(K)	σ(nm)	Ref.
R11	363.6	0.5447	Estimated	ammonia	386	0.2957	[40]
R12	297.24	0.5186	Estimated	argon	143.2	0.335	[41]
R13	204.00	0.4971	[42]	butane	280.51	0.573	[43]
R14	164.44	0.4543	[44]	CO	91.7	0.369	[12]
R32	289.65	0.4098	Fit to [45]	CO2	251.196	0.3751	[46]
R115	201.90	0.5876	[47]	cyclohexane	297.1	0.6182	[43]
R116	226.16	0.5249	Estimated	H2O,D2O	809.1	0.2641	[43]
R124	275.80	0.5501	[48]	decane	490.51	0.686	[49]
R125	249.00	0.5190	[50]	dimethyl ether	395	0.4307	[43]
R141b	370.44	0.5493	Estimated	ethane	245	0.4362	[51]
R142b	278.20	0.5362	[52]	ethanol	362.6	0.453	[43]
R218	266.35	0.5800	Estimated	ethylene	224.7	0.4163	[43]
R227ea	289.34	0.5746	Estimated	H2S	301.1	0.36237	[43]
R236ea	318.33	0.5604	Estimated	helium	10.22	0.2551	[43]
R236fa	307.24	0.5644	Estimated	heptane	400	0.64947	[12]
R245ca	345.44	0.5505	Estimated	hexane	399.3	0.5949	[43]
R245fa	329.72	0.5529	Estimated	hydrogen	59.7	0.2827	[43]
RC318	299.76	0.5947	Estimated	isobutene	332	0.5026	Estimated
R113	376.035	0.6019	Estimated	isohexane	395.2	0.5799	Estimated
Krypton	178.9	0.3655	[43]	isopentane	341.06	0.56232	[12]
Methane	174	0.36652	[53]	isobutane	307.55	0.46445	[54]
Methanol	481.8	0.3626	[43]	nitrogen	98.94	0.3656	[41]
N2O	232.4	0.3828	[43]	nonane	472.127	0.66383	[49]
Neon	32.8	0.282	[43]	oxygen	118.5	0.3428	[41]
neopentane	191	0.644	[55]	propane	263.88	0.49748	[56]
Octane	452.09	0.63617	[49]	R23	243.91	0.4278	[57]
Pentane	341.1	0.5784	[12]	R41	244.88	0.4123	Estimated
Propylene	298.9	0.4678	[43]	R114	323.26	0.5770	Estimated
R22	284.724	0.4666	fit to [58]				

Table 3. Summary of results for the viscosity of binary mixtures for selected systems.

System	T range, K	p range, kPa	x_1 range	1st author	npts	AAD	BIAS	RMS	Max
R134a/R32	298-423	100-7563	0.25-0.75	Yokoyama[59]	329	0.57	0.21	0.74	3.74
R134a/R32	246-335	173-2638	0.3-0.7	Laesecke[28]	148	1.19	0.18	1.37	-3.29
R134a/R32	252-305	256-1347	0.51-0.52	Ripple[29]	15	1.28	-0.14	1.51	-2.82
R134a/R32	223-333	50-3291	0.24-0.76	Heide[26]	21	9.58	9.58	2.91	15.03
R134a/R143a	223-333	45-2546	0.25-0.74	Heide[26]	21	3.24	3.19	1.46	5.63
R134a/R152a	223-333	27-1617	0.25-0.74	Heide[26]	21	3.42	3.42	2.20	7.85
R32/R124	253-311	268-1441	0.46-0.49	Ripple[29]	14	0.96	-0.06	1.12	-1.96
R125/propane	298-423	101-6710	0.50-0.75	Yokoyama[30]	246	0.43	-0.11	0.54	-1.92
R125/R32	298-423	101-7721	0.25-0.74	Yokoyama[60]	357	2.01	2.00	1.07	4.90
R125/R32	223-333	100-3710	0.26-0.75	Heide [26]	21	4.20	1.12	5.47	-13.22
R125/R134a	298-423	101-6624	0.26-0.75	Yokoyama[33]	340	0.54	0.15	1.35	-15.41
R125/R134a	249-345	171-2811	0.3-0.7	Laesecke[28]	156	1.17	0.50	1.29	-4.03
R125/R134a	252-311	224-1408	0.48-0.49	Ripple[29]	15	1.17	-0.06	1.41	-2.87
R125/R143a	223-333	87-3059	0.24-0.75	Heide[26]	42	16.07	-11.34	16.37	-39.76
R125/R152a	223-333	47-2759	0.24-0.75	Heide[26]	21	1.84	0.74	2.43	-6.69
R22/propane	298-348	101	0-1	Nagaoka[31]	21	0.99	-0.30	1.13	-2.35
R115/propane	298-348	101	0-1	Nagaoka[31]	21	0.89	-0.18	1.04	-1.87
R12/R22	298-323	101	0-1	Nagaoka[31]	11	2.90	-2.90	0.93	-4.00
R13/R22	298-323	101	0-1	Nagaoka[31]	11	0.50	-0.15	0.64	-1.51
R14/R22	298-323	101	0-1	Nagaoka[31]	14	0.55	-0.26	0.68	-1.51
R22/R152a	298-323	101	0-1	Nagaoka[31]	10	1.35	-0.90	1.71	-4.51
R125/R143A(R507A)	300-421	90-67754	0.41	Laesecke[27]	145	1.00	-0.02	1.48	-9.80
R32/R125 (R410A)	340-421	90-64452	0.70	Laesecke[27]	145	4.34	-2.32	7.05	-23.06
R32/propane	246-335	434-3930	0.3-0.7	Laesecke[28]	128	1.65	-1.57	1.07	-4.34
R134a/propane	246-345	221-3214	0.35-0.65	Laesecke[28]	162	3.72	2.36	3.38	-11.02
methane/ethane	298-468	101	0.26-0.75	Abe[37]	15	0.36	-0.20	0.40	-1.00
methane/propane	298-498	1820-34528	0.2	Bicher[61]	48	3.18	-0.39	4.15	-11.50
methane/propane	298-468	101	0.28-0.78	Abe[37]	15	0.38	-0.27	0.37	-0.87
methane/butane	298-468	101	0.16-0.74	Abe[37]	15	0.50	-0.27	0.55	-1.10
ethane/propane	298-468	101	0.34-0.86	Abe[37]	17	0.28	0.03	0.38	0.98
ethane/butane	298-468	101	0.19-0.84	Abe[37]	15	0.30	0.00	0.41	-1.02

System	T range, K	p range, kPa	x_1 range	1st author	npts	AAD	BIAS	RMS	Max
propane/butane	298-468	101	0.22-0.80	Abe[37]	19	1.12	-0.01	2.32	-9.55
methane/CO2	273	104-2572	0.14-0.67	Kestin[36]	32	0.17	-0.10	0.17	-0.51
methane/butane	273	105-668	0.36-0.84	Kestin[36]	34	0.40	0.19	0.46	1.14
methane/nitrogen	100-300	1603-33866	0.31-0.71	Diller[38]	306	1.73	0.18	2.21	-7.53
CO/krypton	298-473	101	0.42-0.66	Kestin[32]	10	0.40	-0.08	0.42	-0.67
CO/CO2	298-473	101	0.23-0.68	Kestin[32]	10	0.28	-0.15	0.29	-0.54
CO/methane	298-473	101	0.33-0.68	Kestin[32]	10	0.26	-0.20	0.23	-0.56
CO/R14	298-473	101	0.32-0.60	Kestin[32]	10	0.33	-0.09	0.36	-0.63
methane/hydrogen	173-273	404-50586	0.21-0.81	Chuang[62]	145	1.93	-0.21	2.31	-4.68
decane/dodecane	298	101	0-1	Aucejo[25]	11	0.56	-0.14	0.68	-1.47
nonane/dodecane	298	101	0-1	Aucejo[25]	11	0.60	-0.22	0.70	-1.47
nonane/decane	298	101	0-1	Aucejo[25]	11	0.53	-0.07	0.63	-1.23
octane/dodecane	298	101	0-1	Aucejo[25]	11	1.21	-0.20	1.31	2.01
octane/decane	298	101	0-1	Aucejo[25]	11	0.54	-0.06	0.65	-1.23
octane/nonane	298	101	0-1	Aucejo[25]	11	1.13	-0.80	1.04	-2.17
heptane/dodecane	298	101	0-1	Aucejo[25]	11	1.85	-0.32	2.07	3.29
heptane/decane	298	101	0-1	Aucejo[25]	11	0.88	-0.09	1.10	-2.02
heptane/nonane	298	101	0-1	Aucejo[25]	11	1.52	-1.40	1.18	-2.95
heptane/octane	298	101	0-1	Aucejo[25]	11	0.39	0.07	0.46	0.86
hexane/dodecane	298	101	0-1	Aucejo[25]	11	3.07	-0.80	3.31	5.14
hexane/decane	298	101	0-1	Aucejo[25]	11	1.19	-0.3	1.54	-2.87
hexane/nonane	298	101	0-1	Aucejo[25]	11	0.89	-0.29	0.98	-2.09
hexane/octane	298	101	0-1	Aucejo[25]	11	0.49	-0.08	0.60	-1.12
hexane/heptane	298	101	0-1	Aucejo[25]	11	0.53	-0.39	0.46	-1.06
pentane/dodecane	298	101	0-1	Aucejo[25]	11	5.98	-2.02	6.39	-9.74
pentane/decane	298	101	0-1	Aucejo[25]	11	2.99	-0.91	3.25	-5.46
pentane/nonane	298	101	0-1	Aucejo[25]	11	1.88	-0.67	1.93	-2.89
pentane/octane	298	101	0-1	Aucejo[25]	11	0.72	-0.26	0.90	-2.29
pentane/heptane	298	101	0-1	Aucejo[25]	11	1.21	0.34	1.29	-2.29
pentane/hexane	298	101	0-1	Aucejo[25]	11	1.97	-1.97	0.54	-2.64
decane/CO2	311-403	6720-34680	0.15-0.85	Cullick[23]	83	21.7	-8.01	24.10	70.65
methane/propane	311-411	101-55158	0.22-0.79	Giddings[63]	282	2.00	-1.58	2.39	17.74
pentane/heptane	443-526	1243-3453	0.26-0.74	Abdulagatov[64]	70	11.09	-7.75	12.86	-43.21

System	T range, K	p range, kPa	x_1 range	1st author	npts	AAD	BIAS	RMS	Max
cyclohexane/hexane	298	101	0-1	Tripathi[39]	15	1.58	0.08	1.74	2.99
decane/hexane	298	101	0-1	Tripathi[39]	15	5.99	5.58	4.95	14.35
hydrogen/nitrogen	195-523	101	0-1	Trautz[34]	36	1.11	0.21	1.60	6.81
hydrogen/CO	195-523	101	0-1	Trautz[34]	36	0.99	0.51	1.39	5.12
hydrogen/neon	293-523	101	0-1	Trautz[34]	24	1.29	0.48	1.41	2.80
hydrogen/helium	293-523	101	0-1	Trautz[34]	20	0.73	0.67	0.46	1.36
hydrogen/argon	293-523	101	0-1	Trautz[34]	24	0.76	0.40	0.80	2.24
hydrogen/ammonia	293-523	101	0-1	Trautz[34]	31	1.51	0.18	1.81	-4.10
hydrogen/methane	293-523	101	0-1	Trautz[34]	24	1.03	-0.29	1.23	-2.85
hydrogen/ethane	293-523	101	0-1	Trautz[34]	24	1.62	-0.39	1.89	-3.58
hydrogen/propane	300-550	101	0-1	Trautz[34]	32	1.92	-0.18	2.32	-4.40
hydrogen/CO2	300-550	101	0-1	Trautz[34]	32	1.12	0.30	1.31	-2.67
hydrogen/oxygen	300-550	101	0-1	Trautz[34]	32	0.84	-0.04	0.99	-2.12
hydrogen/ethylene	195-323	101	0-1	Trautz[34]	50	1.97	0.09	3.12	10.68
hydrogen/propylene	293-523	101	0-1	Trautz[34]	44	2.02	-0.88	2.30	-5.08
hydrogen/cis-2-butene	293-523	101	0-1	Trautz[34]	40	2.52	1.60	2.91	7.67

Table 4. Summary of results for the viscosity of multicomponent mixtures.

System	T range, K	p range, kPa	$x_{methane}$	1st author	npts	AAD	BIAS	RMS	Max
refrigerant mixture (R32/R125/R134A)	246-340	221-2853	na	Laesecke[28]	157	1.39	1.28	1.12	3.88
natural gas H	260-320	95-20073	0.896	Schley[65]	224	0.71	-0.71	0.54	-2.26
natural gas L	260-320	95-20227	0.843	Schley[65]	224	1.20	-1.20	0.30	-1.62
natural gas 1	298-339	101-69154	0.735	Carr[66]	53	1.42	0.45	2.37	10.71
natural gas 2	299-394	101-66051	0.731	Carr[66]	35	2.38	-2.00	1.55	-4.31
natural gas 3	303-398	101-58364	0.956	Carr[66]	33	1.01	-0.32	1.14	-2.47
natural gas	241-455	242-14040	0.848	Assael[67]	40	1.61	-1.60	1.00	-4.47
natural gas 1	263-303	4460-25110	0.902	Langelandsvik[68]	45	5.05	-5.05	0.96	-6.69
natural gas 2	263-304	13310-25310	0.800	Langelandsvik[68]	34	6.18	-6.18	1.07	-8.50
natural gas 3	263-304	4950-25240	0.922	Langelandsvik[68]	45	4.30	-4.30	1.09	-6.75

We also made comparisons of the viscosity prediction of the REFPROP model for multicomponent mixtures. Comparisons with a ternary refrigerant mixture R32/R125/R134A measured by Laesecke et al. [28] are presented in Table 4 and Figures 1a-c. In the Figures, the percent deviation is defined as 100*(calculated value-experimental)/experimental. Two different compositions 0.33/0.33/0.34 and 0.3/0.1/0.6 mole fraction of R32/R125/R134a were measured from 246 to 340 K at pressures up to 2853 kPa with average absolute deviations of less than two percent. The estimated uncertainty of this data is given by the author as 2.4 %. We also made comparisons with several natural gas mixtures. Figures 2a-c show the deviations between the REFPROP model and experimental data for several different natural gas mixtures as a function of density, pressure and temperature. Schley et al. [65] measured two different natural gases at pressures to 20 MPa. The "L" gas is a low calorific gas with a relatively high content (9.75 mole %) of nitrogen; the high calorific "H" natural gas contains only 1.5 % nitrogen. Both samples are represented well by the REFPROP model. We note that a binary interaction parameter for methane/nitrogen viscosity was added to the REFPROP program after the initial release of Version 8 in April 2007; this parameter significantly improved the results for the high-nitrogen natural gas. Carr [66] measured three natural gas samples. Sample 1 is a high-ethane natural gas (73.5 mole % methane, 25.7 % ethane, 0.6 % N_2 and 0.2 % propane), Sample 2 is a high-nitrogen natural gas (73.1 % methane, 15.8 % N_2, 6.1 % ethane, 3.4% propane, 0.8 % helium, 0.6 % n-butane and 0.2 % isobutane) obtained from a transmission line, and Sample 3 is a simulated low-ethane natural gas (95.6 % methane, 3.6 % ethane, 0.5 % propane and 0.3 % N_2). With the exception of the Sample 1 gas at pressures above 20 MPa, the model represents the data well. The Assael et al. [67] natural gas is a 5 component mixture of methane, ethane, propane, CO_2 and nitrogen (84.8, 8.4, 0.5, 5.6, and 0.66 mol % respectively) at pressures up to 14 MPa and is represented with an average absolute deviation of less than two percent, but shows systematic deviations with the largest deviations at the lowest temperatures. Systematic deviations are even more pronounced in the data from Langelandsvik et al. [68] who measured the viscosity of natural gas samples from three different locations in the North Sea. The primary constituents in gas 1 are: methane 90.2 %, ethane 6.3 %, propane 0.8 % and CO_2 1.8 %; in gas 2: methane 80.0 %, ethane 9.3 %, propane, 5.0 %, CO_2 2.2 %; and in gas 3: methane 92.2 %, ethane 4.4 %, propane 0.5 %, CO_2 1 %. REFPROP shows systematic negative deviations for all the Langelandsvik et al. [68] samples with the deviations increasing as the pressure increases at the lowest temperatures. It was noted in the Langelandsvik et al. [68] manuscript that there are unexplained differences between the measurements of Schley et al. [65] and the work of Langelandsvik et al. [68]. The REFPROP program is in better agreement with the measurements of Schley et al. [65].

The thermal conductivity model was tested on a limited set of thermal conductivity mixture data containing both refrigerant mixtures and hydrocarbon mixtures. When possible, we obtained binary interaction parameters to improve the representation of the data. The resulting values for the binary interaction parameters for both the dilute and residual thermal conductivity are given in Table 5 below.

Table 5. Values of interaction parameters for thermal conductivity for selected systems.

System	$k_{ij,\lambda}$	$\lambda_{ij,f\lambda}$	$k_{ij,h\lambda}$	Data source
methane/CO2 -0.15		0.0	0.0	[69] [70]
methane/CO -0.07		0.0	0.0	[71]
methane/propane -0.2		0.0	0.0	[72]
methane/ethane -0.12		0.0	0.0	[73]
R22/R142b -0.76		0.0	0.0	[74] [75]
R32/propane 0.18		-0.08	0.09	[76]
R32/R134a 0.08		-0.09	0.1	[76]
R32/R125	0.0	-0.15	0.16	[77, 78] [79]
R134a/propane 0.07		-0.04	0.20	[76]

Comparisons for both binary and multicomponent mixtures were made. The results are summarized in Tables 6-8 below. Most systems that do not contain data near the critical region have deviations of less than 5 % with the notable exception of the binary system R32/R125 measured by Kim et al. [80], which shows deviations over 10 %. Since this same binary system was measured by other researchers [77-79] and the model agrees well with the experimental data, there must be an unexplained problem with the data of Kim et al. [80] for this binary system. Perkins et al. [76] made extensive measurements on binary mixtures containing R32, R125, R134a, and propane. The measurements covered the gas, liquid and some near-critical points. Figures 3a-d show the deviations for Perkins et al. [76] measurements as a function of density. The REFPROP model represents the data well except for points where the critical enhancement is significant; in these cases the model significantly underestimates the thermal conductivity.

Roder and Friend [73] made extensive measurements on the thermal conductivity of a binary mixture of methane and ethane for three different concentrations (0.7, 0.5, and 0.35

mol fraction methane) over a range of conditions that included regions where the enhancement is significant. Sakonidou et al. [21] also measured the methane/ethane system in the critical region. Figures 4a-d show the percent deviations of the data from the model with and without the critical enhancement. The figure demonstrates the improvement seen when the enhancement term is included; however, it also demonstrates that the present model is not adequate, and further work in this area is needed.

5. Conclusions

We describe a modified corresponding states model based on an extended corresponding states model originally developed by Ely and Hanley [2, 3] for the viscosity and thermal conductivity of mixtures. The model has been implemented in the computer program REFPROP [1], version 8, available from NIST. Comparisons with selected mixture data for viscosity and thermal conductivity are presented. The model represents the viscosity and thermal conductivity of mixtures to within 5 to 10 %, except for mixtures containing large size differences, where the deviations are larger. A model for the thermal conductivity of mixtures in the critical region is also implemented, and although it provides some enhancement to the thermal conductivity, comparisons with experimental data indicate that the model significantly underpredicts the thermal conductivity in the critical region, and further research in this area is recommended.

We thank our NIST colleagues Dr. Richard Perkins and Dr. Daniel Friend for their helpful comments. JC acknowledges a Professional Research Experiences Program (PREP) undergraduate fellowship at NIST.

Table 6. Summary of results for the thermal conductivity of binary mixtures for selected systems.

System	T range, K	p range kPa	x_1 range	1st Author	npts	AAD	BIAS	RMS	Max
methane/CO2	300-425	724-11970	0.25-0.75	Patek [69]	180	1.18	-1.09	1.07	-4.50
methane/CO2	298-398	100-9000	0.12-0.93	Yorizane[81]	62	2.10	0.33	2.61	-6.29
methane/CO2	300-301	852-6300	0.26-0.79	Kestin[70]	96	1.52	-1.52	0.67	-4.11
methane/CO	296.73-300.65	866.7-12200	0.25-0.77	Imaishi[71]	82	1.27	-0.55	1.51	4.24
methane/nitrogen	299.51-300.65	854-8280	0.26-0.77	Kestin [82]	110	0.75	-0.67	0.53	-1.44
methane/propane	323.15-423.15	101	0-1	Smith[72]	32	2.42	0.93	3.00	8.28
R22/R142b	164.77-295.76	2700-8140	0.41-0.79	Tsvetkov[74]	29	1.70	0.55	1.92	4.14
R22/R142b	223.15-323.15	2100-20100	0-1	Kim[75]	125	0.64	-0.03	0.80	1.79
R22/R152a	176.6-297.45	2440-8020	0.22-0.66	Tsvetkov[74]	13	2.83	2.53	2.11	6.96
R22/R152a	223.15-323.15	2100-20100	0-1	Kim[75]	125	1.37	-0.32	1.77	-4.69
R32/R125	223.15-323.16	2000-20000	0-1	Kim[80]	120	12.27	11.35	11.38	37.08
R32/R125	231.25-324.05	2000-20001	0-1	Ro[77]	168	1.59	-0.39	1.83	-4.55
R32/R125	283.15-298.15	100-1200	0.19-0.82	Tanaka[78]	69	1.67	-0.78	1.93	-4.66
R32/R125	213-293	2000-30000	0.43-0.87	Gao[79]	60	1.85	1.20	1.79	-6.36
R32/R134a	223.15-323.15	2000-25000	0-1	Ro[83]	120	2.01	1.09	2.22	5.05
R32/R134a	193.2-316.1	2000-30000	0.4-0.85	Gao[79]	84	1.40	1.16	1.24	4.29
R32/R134a	253.71-360.68	83-11694	0.3-0.7	Perkins[76]	1308	1.29	-0.52	1.58	-6.92
R32/propane†	227.87-346.93	24-10758	0.3-0.7	Perkins[76]	1259	3.92	-0.60	6.49	-52.55
R134a/propane†	242.61-348.50	83-19687	0.3-0.7	Perkins[76]	977	2.02	-0.27	3.65	-35.26
R125/R134a	231.25-324.05	2000-20000	0-1	Jeong[84]	147	0.76	0.36	0.88	-2.64
R125/R134a†	244.30-347.04	70-9949	0.3-0.7	Perkins[76]	1091	2.73	-1.52	2.85	-15.70

† Includes some points showing critical enhancement

Table 7. Summary of results for the thermal conductivity of multicomponent mixtures for selected systems.

System	T range, K	p range, kPa	x_1 range	1st Author	npts	AAD	BIAS	RMS	Max
R32/R125/R134a	232.55-324.15	2000-20000	0.18-0.61	Jeong[85]	168	5.81	-5.81	2.60	-10.03
R32/R125/R134a	248.63-346.96	85-12265	0.3-0.33	Perkins[76] †	1036	2.28	-0.24	2.81	-10.65

† Includes some points showing critical enhancement

Table 8. Summary of results for the thermal conductivity in the critical region.

System	T range, K	p range, kPa	x_1 range	1st Author	npts	AAD	BIAS	RMS	Max
Methane/Ethane	257.44-310.91	180-16152	0.37-0.54	Sakonidou[21]	54	22.43	-22.43	11.97	-47.00
Methane/Ethane	139.71-332.05	228-68255	0.35-0.70	Roder[73]	2476	4.35	-2.92	5.05	-25.56

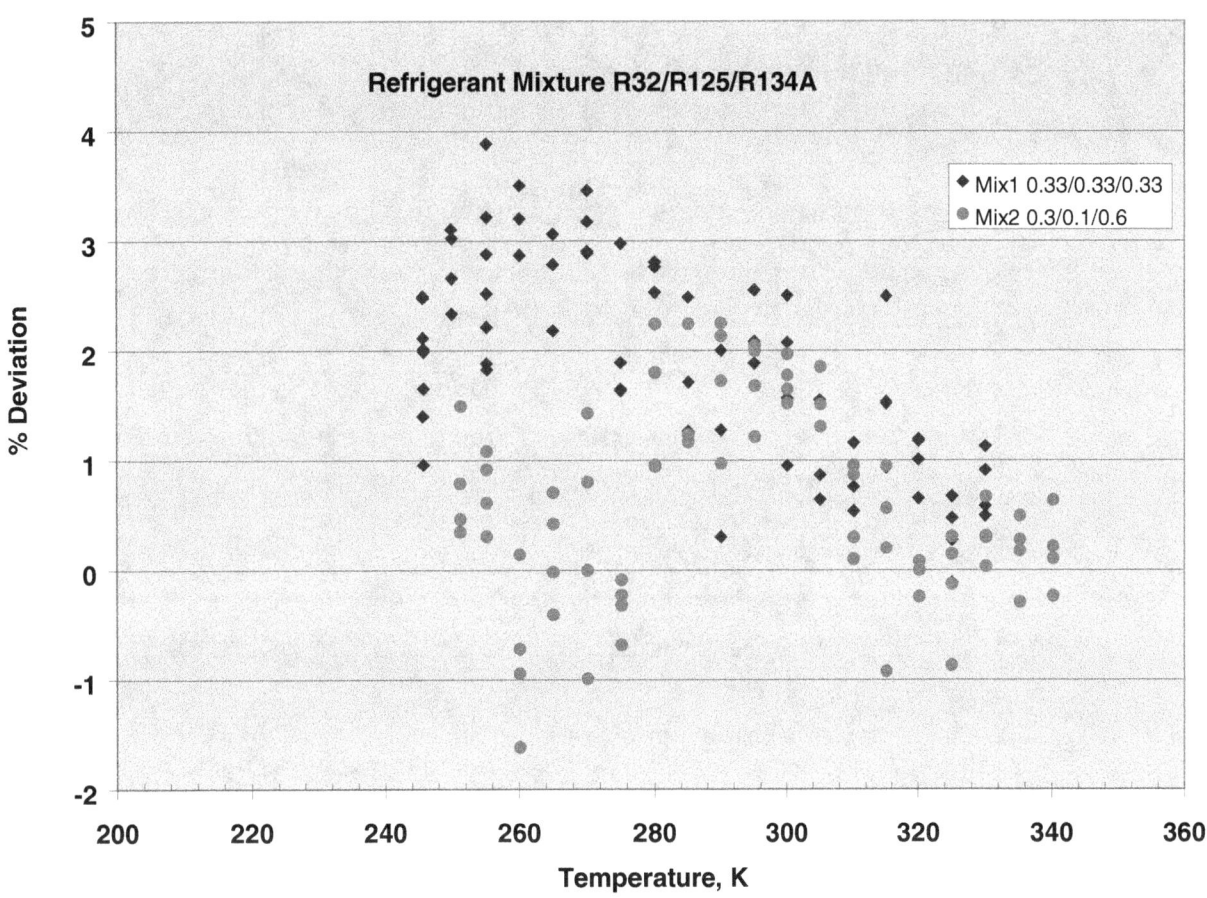

Figure 1a. Percent deviation of viscosity as a function of temperature for a mixture of refrigerants R32, R125 and R134a.

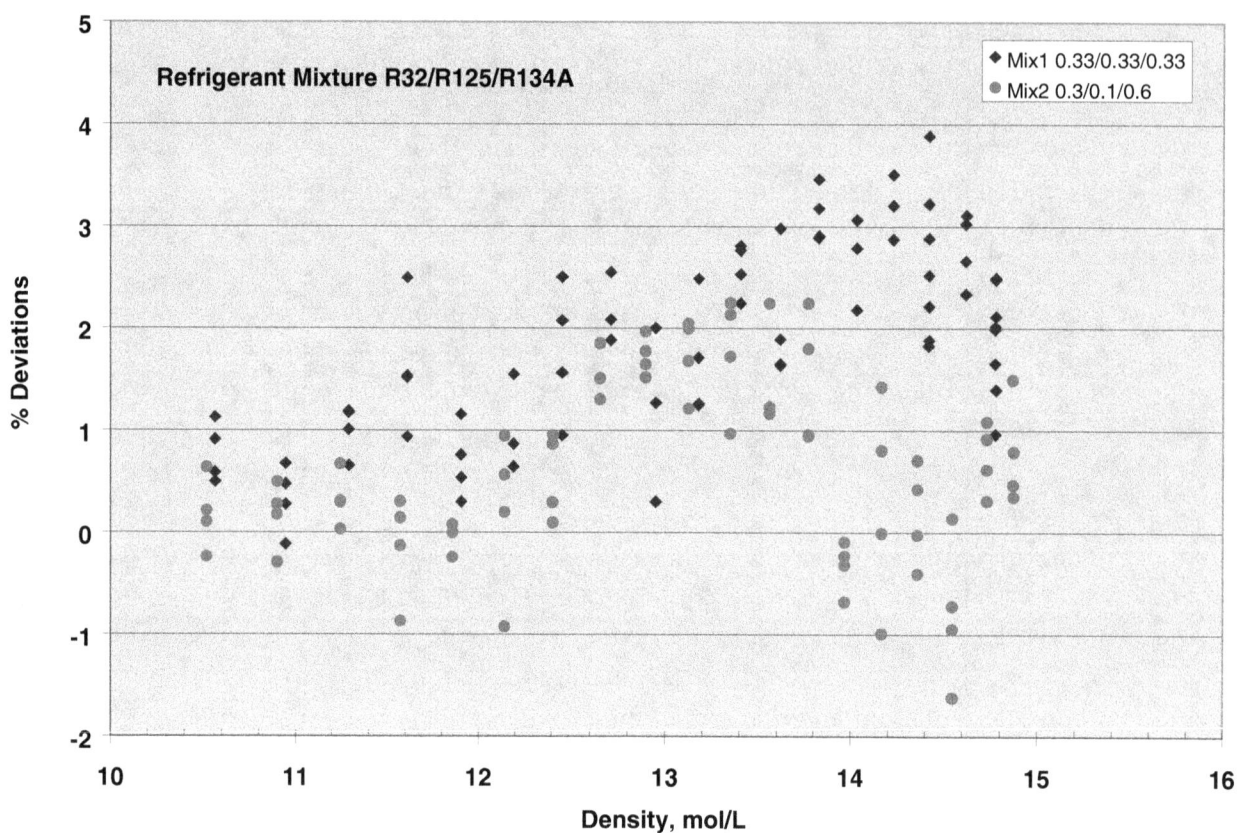

Figure 1b. Percent deviation of viscosity as a function of density for a mixture of refrigerants R32, R125 and R134a.

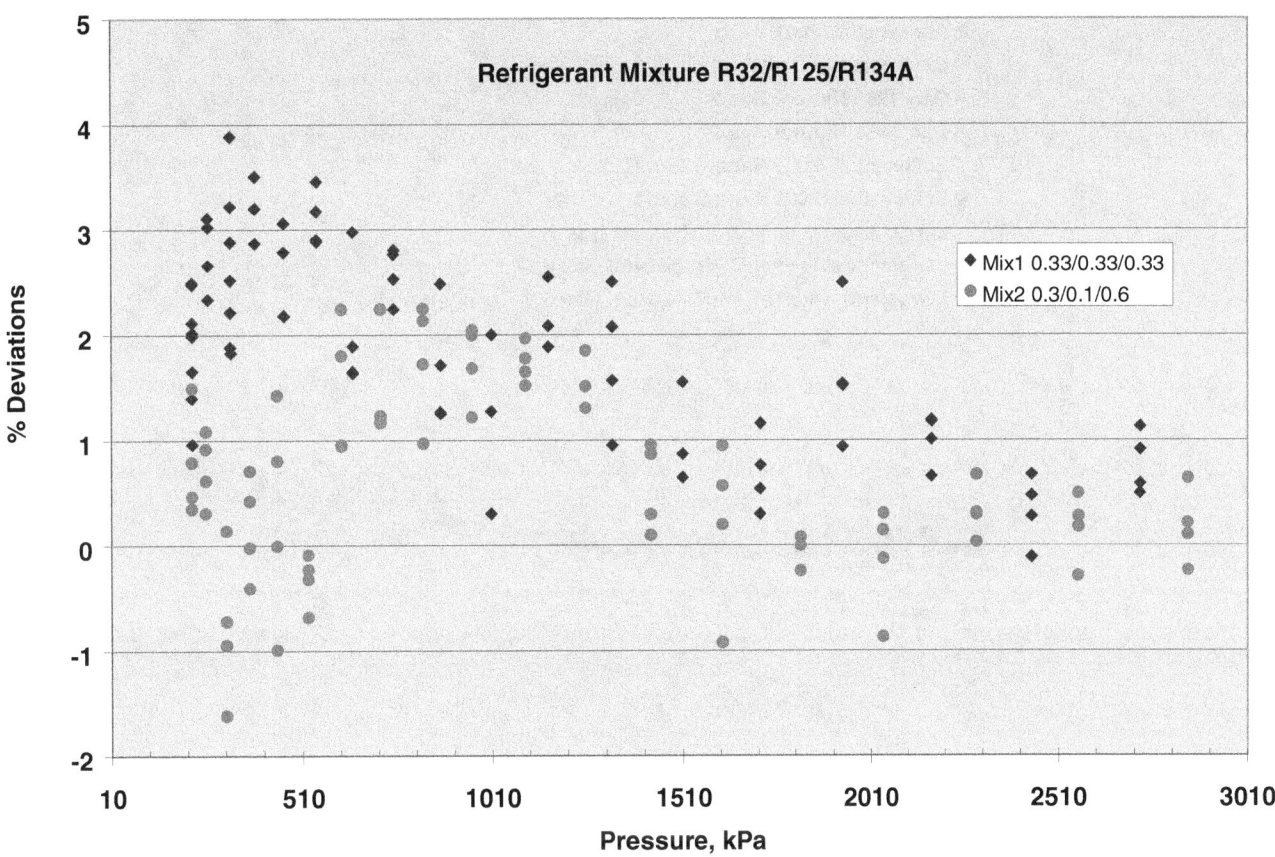

Figure 1c. Percent deviation of viscosity as a function of pressure for a mixture of refrigerants R32, R125 and R134a.

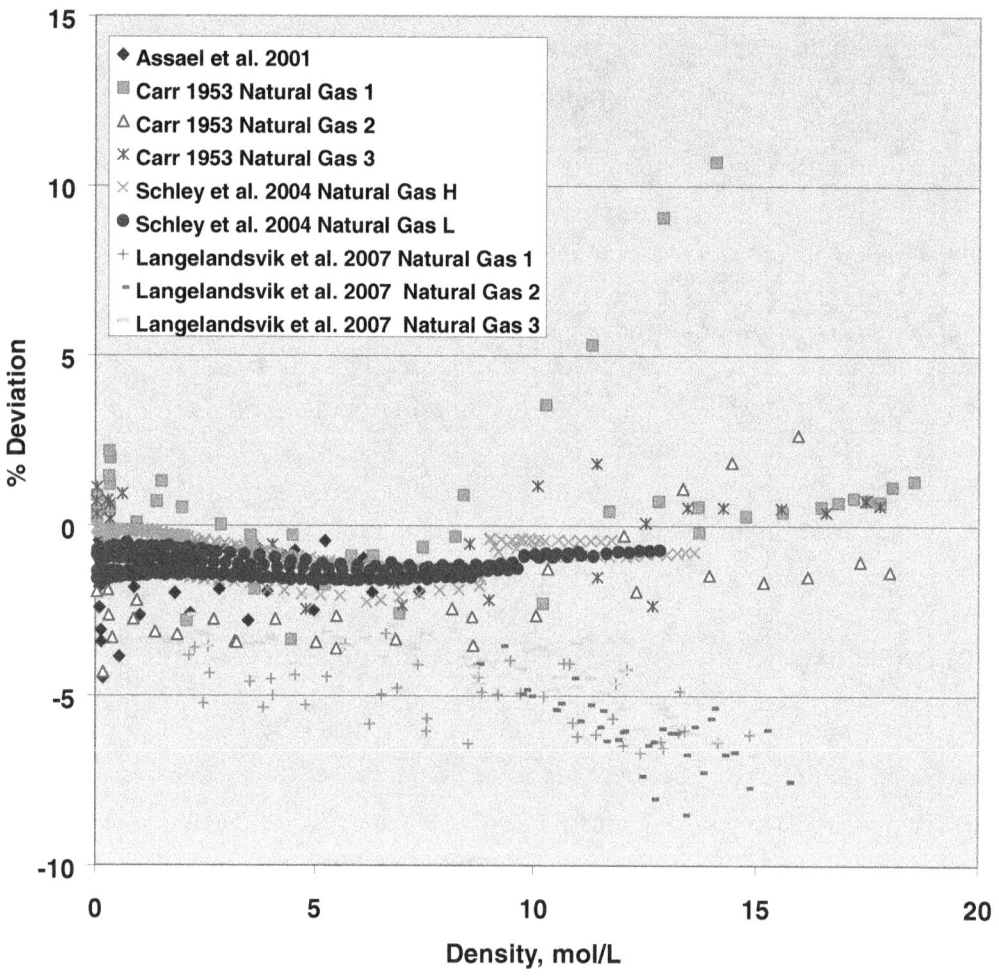

Figure 2a. Percent deviation of viscosity as a function of density for selected natural gas mixtures.

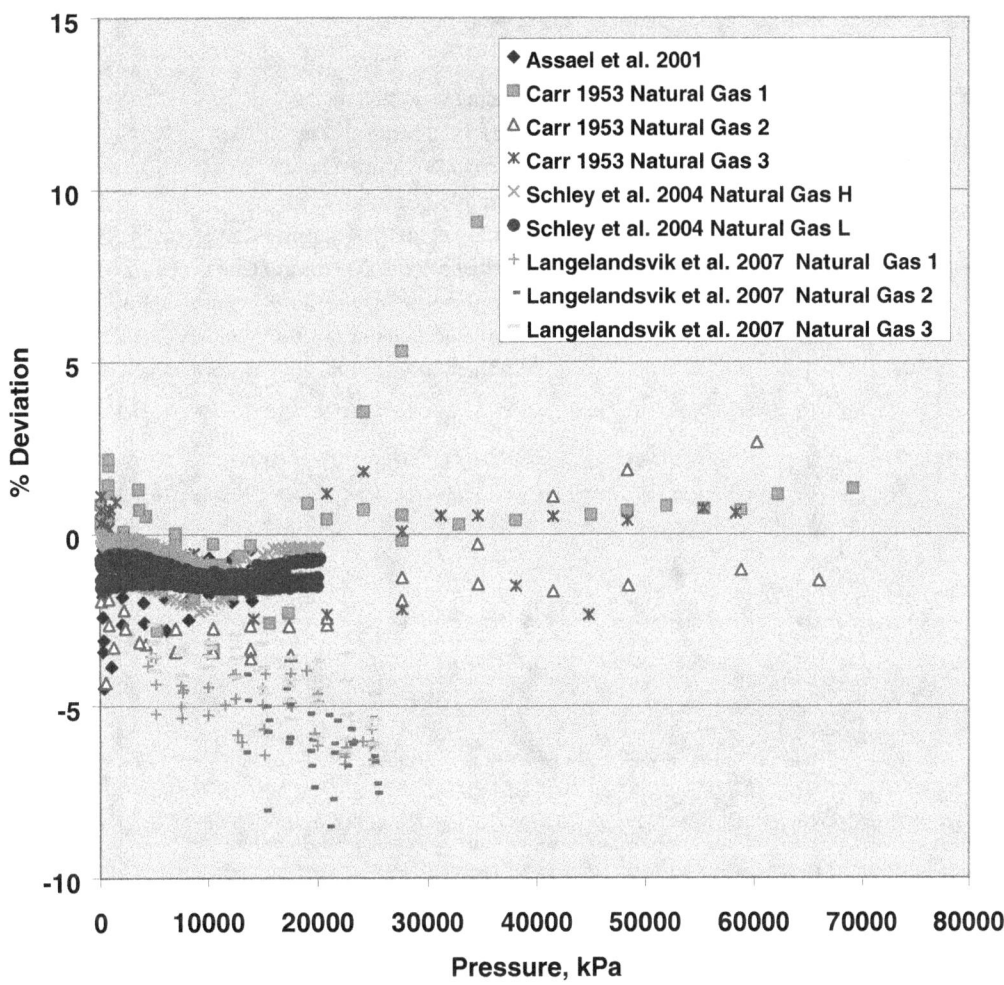

Figure 2b. Percent deviation of viscosity as a function of pressure for selected natural gas mixtures.

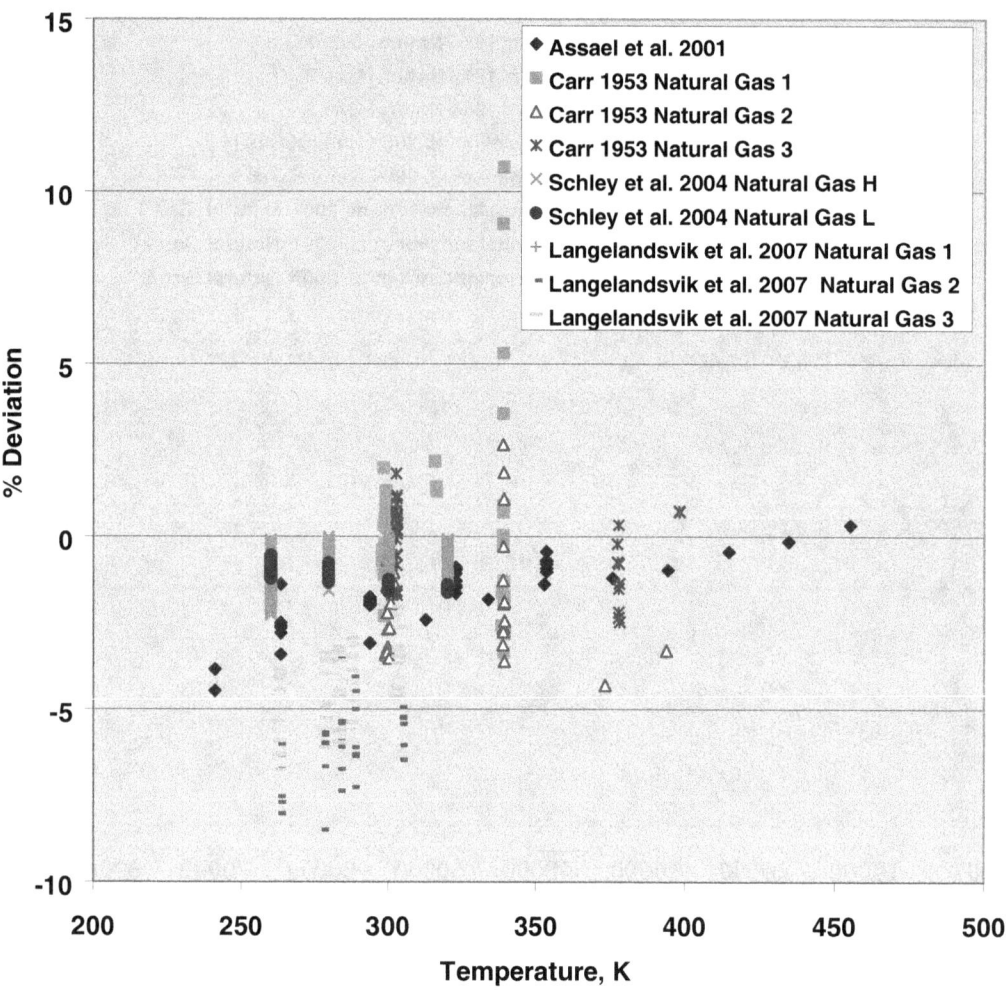

Figure 2c. Percent deviation of viscosity as a function of temperature for selected natural gas mixtures.

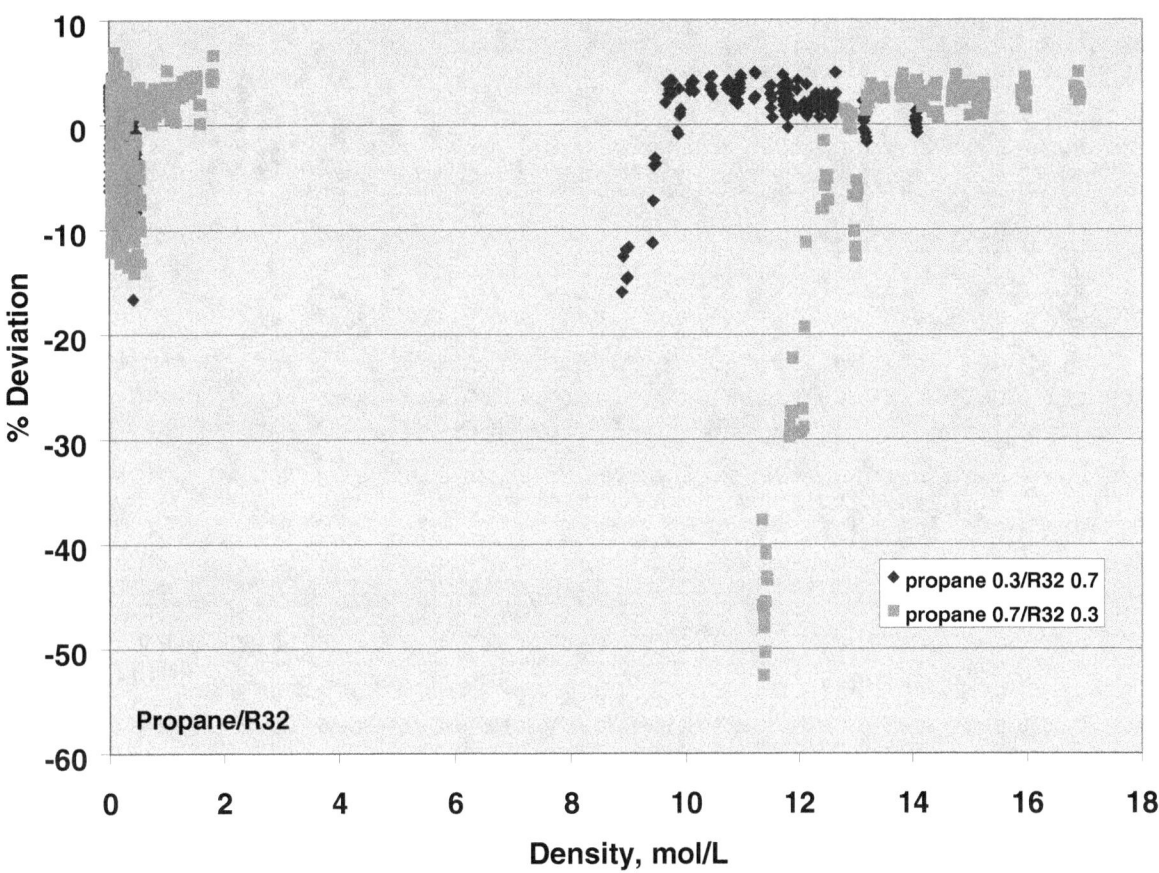

Figure 3a. Percent deviation of thermal conductivity as a function of density for binary mixtures of propane and R32.

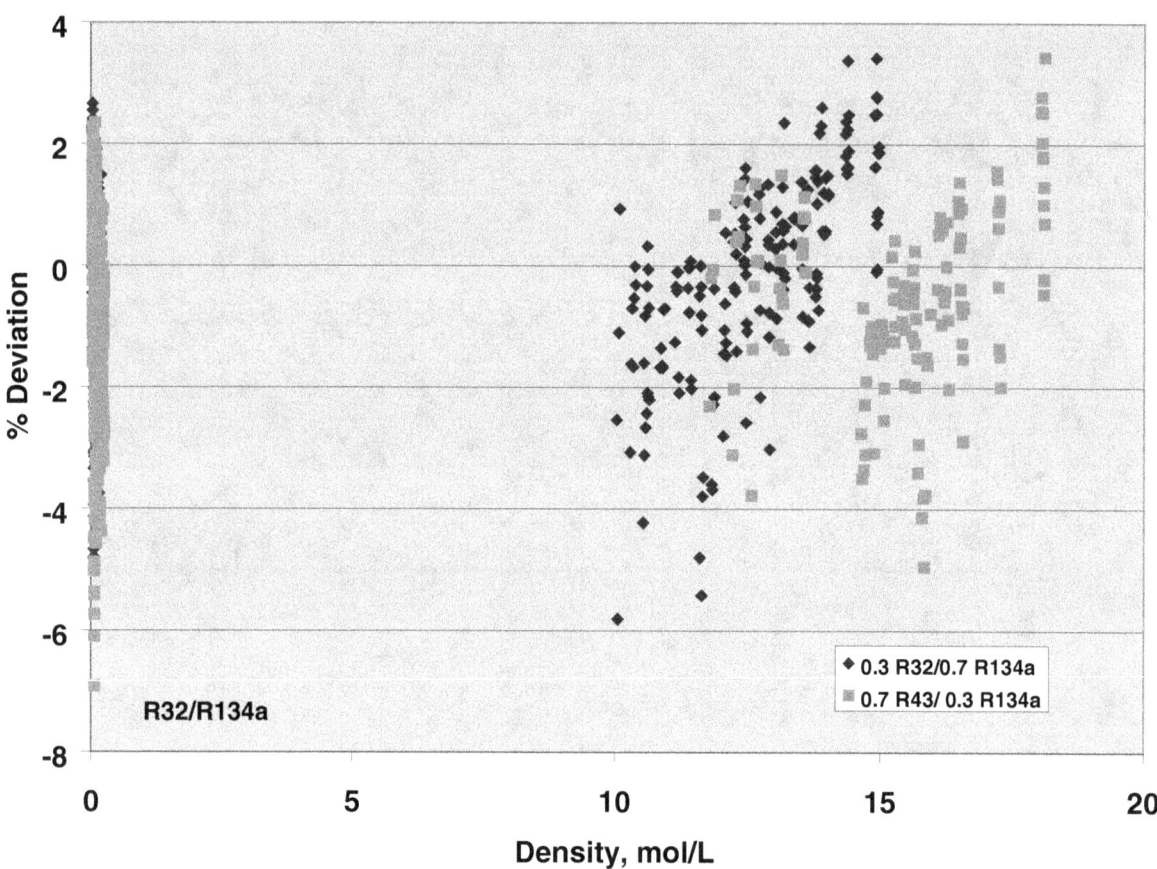

Figure 3b. Percent deviation of thermal conductivity as a function of density for binary mixtures of R134a and R32.

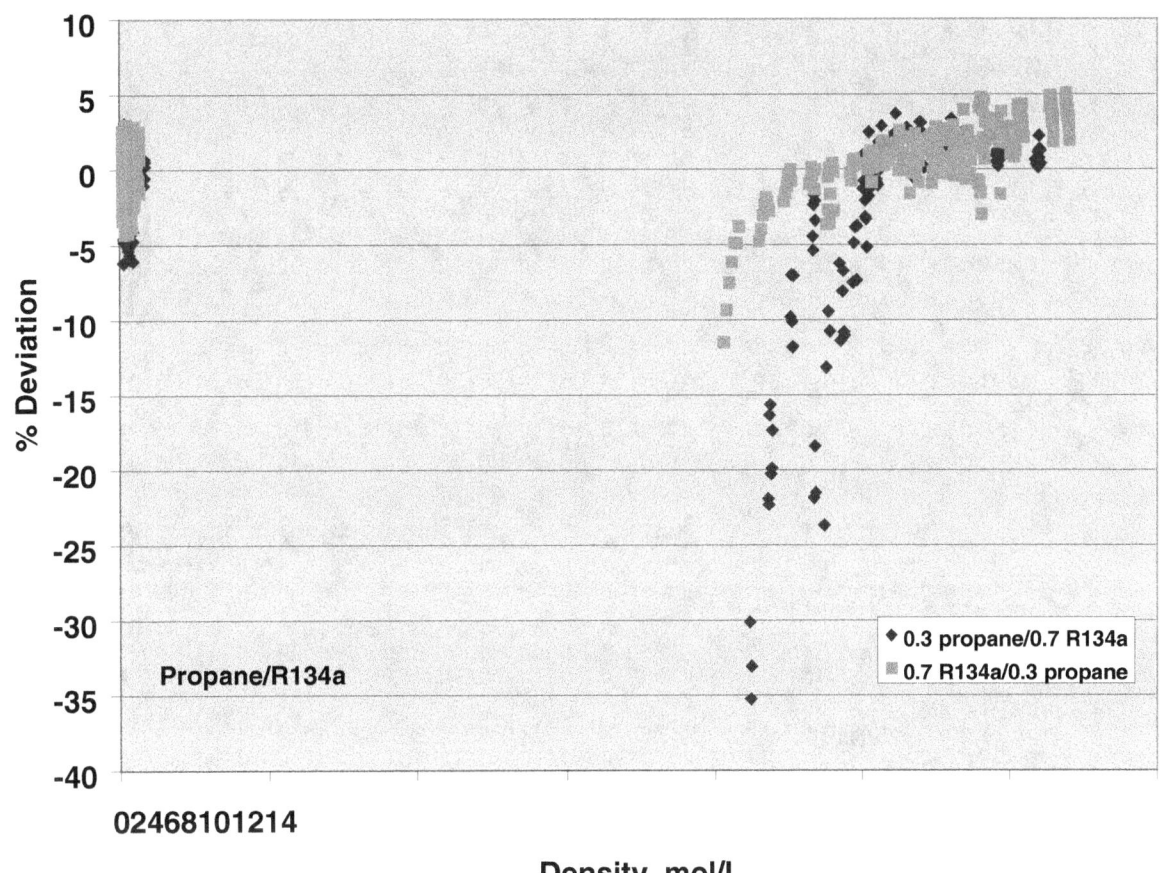

Figure 3c. Percent deviation of thermal conductivity as a function of density for binary mixtures of propane and R134a.

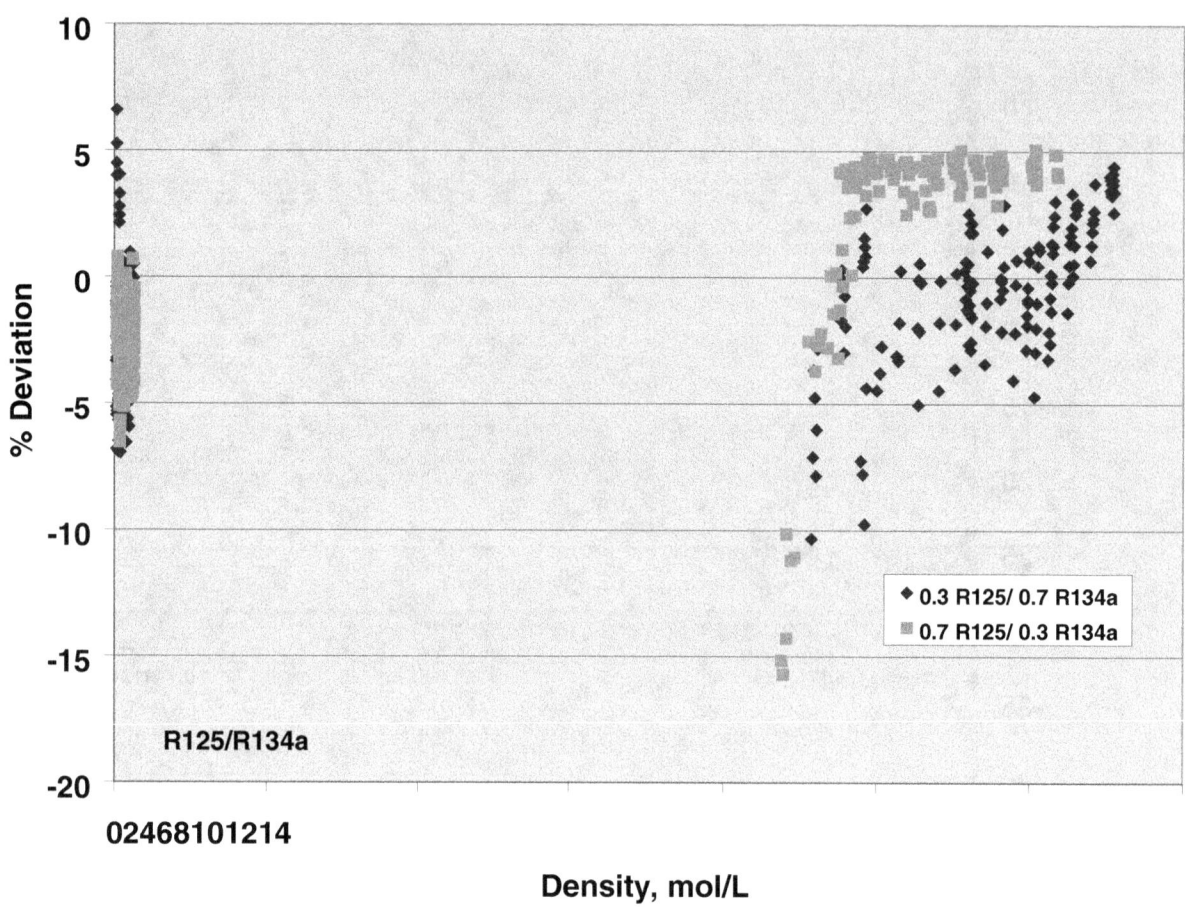

Figure 3d. Percent deviation of thermal conductivity as a function of density for binary mixtures of R125 and R134a.

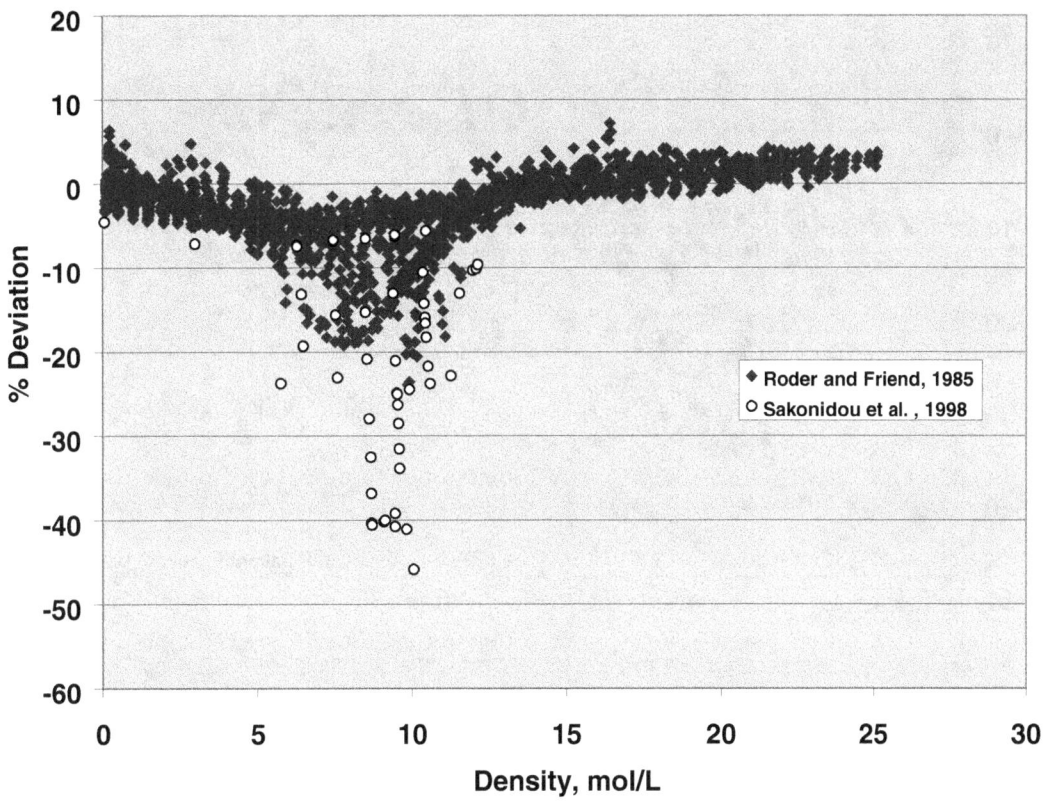

Figure 4a. Percent deviation of thermal conductivity as a function of density for binary mixtures of methane and ethane near the critical region with a critical enhancement term in the model.

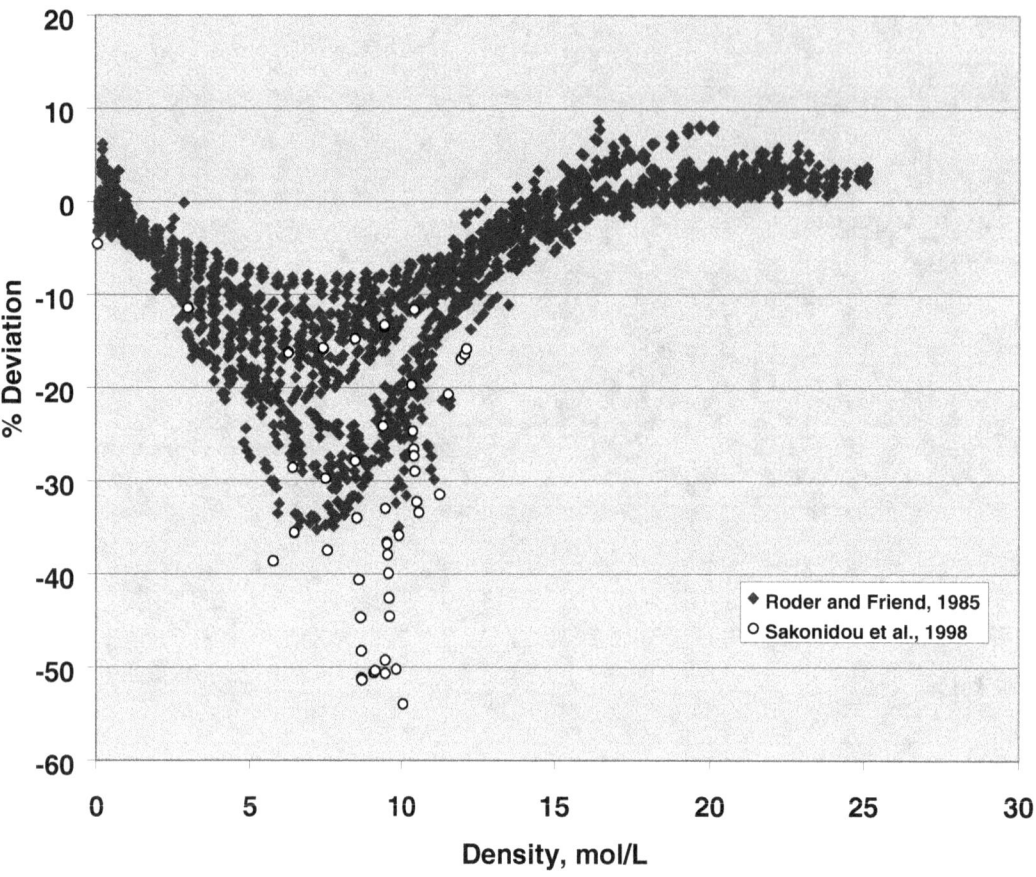

Figure 4b. Percent deviation of thermal conductivity as a function of density for binary mixtures of methane and ethane near the critical region without a critical enhancement term in the model.

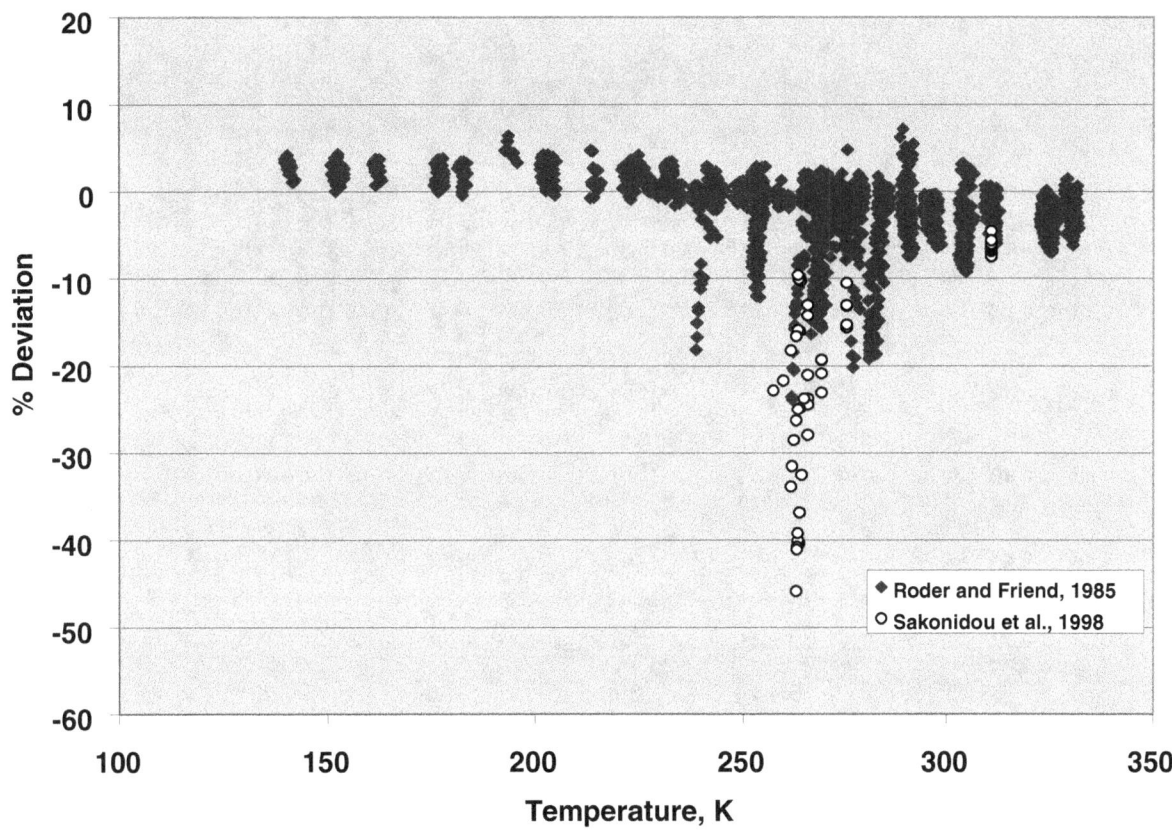

Figure 4c. Percent deviation of thermal conductivity as a function of temperature for binary mixtures of methane and ethane near the critical region with a critical enhancement term in the model.

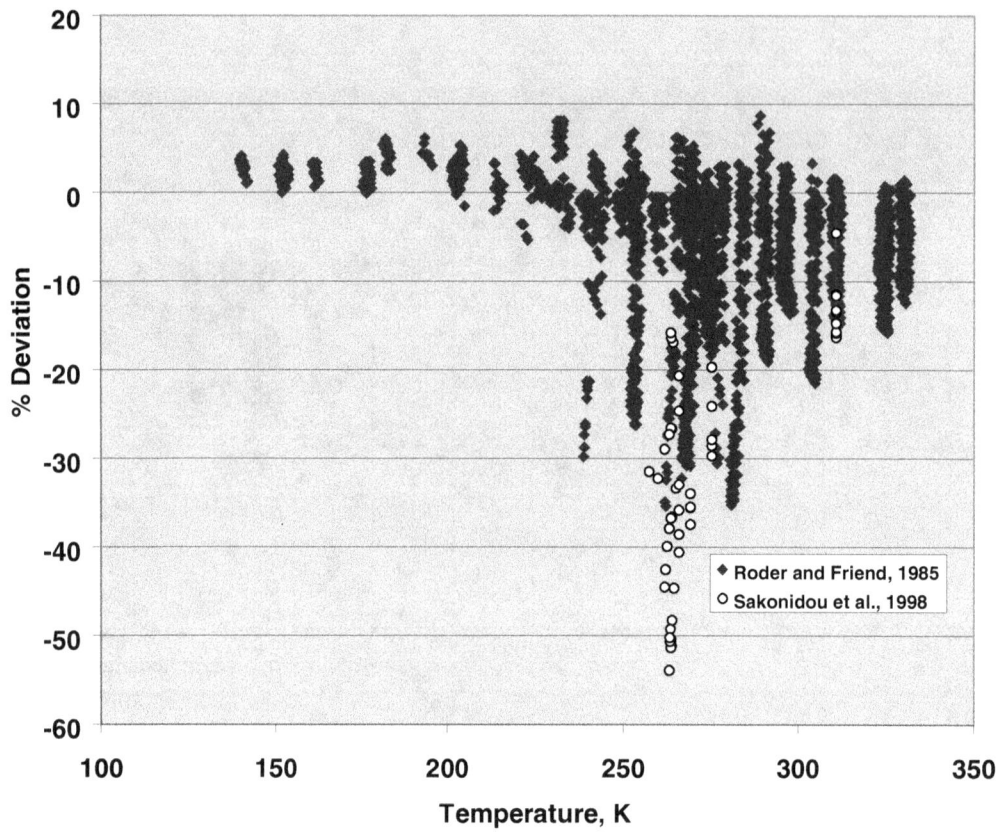

Figure 4d. Percent deviation of thermal conductivity as a function of temperature for binary mixtures of methane and ethane near the critical region without a critical enhancement term in the model.

6. References

1. E. W. Lemmon, M. L. Huber, and M. O. McLinden, NIST reference fluid thermodynamic and transport properties database, NIST 23, (REFPROP), v8.0. 2007, National Institute of Standards and Technology: Gaithersburg, MD.

2. J. F. Ely, and H. J. M. Hanley, Prediction of transport properties. 1. Viscosity of fluids and mixtures. Ind. Eng. Chem. Fundam. 20(4): 323-332 (1981).

3. J. F. Ely, and H. J. M. Hanley, Prediction of transport properties. 2. Thermal conductivity of pure fluids and mixtures. Ind. Eng. Chem. Fundam. 22(1): 90-7 (1983).

4. J. F. Ely, Application of the extended corresponding states model to hydrocarbon mixtures, Proceedings of the 63rd Annual Gas Processors Association Meeting, pp. 9-22 (1984).

5. M. L. Huber, and J. F. Ely, Prediction of the viscosity of refrigerants and refrigerant mixtures. Fluid Phase Equilibria. 80: 239-248 (1992).

6. M. L. Huber, and J. F. Ely, Prediction of the thermal conductivity of refrigerants and refrigerant mixtures. Fluid Phase Equilibria. 80: 249-261 (1992).

7. M. L. Huber, Ely, J.F., A predictive extended corresponding states model for pure and mixed refrigerants including an equation of state for R134a. Int. J. Refrig. 17: 18-31 (1994).

8. M. O. McLinden, S. A. Klein, and R. A. Perkins, An extended corresponding states model for the thermal conductivity of pure refrigerants and refrigerant mixtures. Int. J. Refrig. 23(1): 43-63 (2000).

9. S. A. Klein, M. O. McLinden, and A. Laesecke, An improved extended corresponding states method for estimation of viscosity of pure refrigerants and mixtures. Int. J. Refrig. 20(3): 208-217 (1997).

10. M. L. Huber, A. Laesecke, and R. A. Perkins, Model for the viscosity and thermal conductivity of refrigerants, including a new correlation for the viscosity of R134a. Ind. Eng. Chem. Res. 42: 3163-3178 (2003).

11. M. L. Huber, NIST thermophysical properties of hydrocarbon mixtures, NISTt4 (Supertrapp), v3.1. 2003, Standard Reference Data, National Institute of Standards and Technology: Gaithersburg, MD.

12. D. G. Friend, NIST mixture property database, NIST14 (DDMIX), v9.08. 1992, Standard Reference Data, National Institute of Standards and Technology: Gaithersburg, MD USA.

13. E. W. Lemmon, M. O. McLinden, and M. L. Huber, NIST reference fluid thermodynamic and transport properties database, NIST 23, (REFPROP), v7.0. 2002, National Institute of Standards and Technology: Gaithersburg, MD.

14. J. O. Hirschfelder, C. F. Curtiss, and R. B. Bird, Molecular Theory of Gases and Liquids. Corr. print. with notes attached ed. Structure of Matter Series. 1964, 1954, John Wiley and Sons, Inc., New York.

15. P. D. Neufeld, A. R. Janzen, and R. A. Aziz, Empirical equations to calculate 16 of the transport collision integrals $\omega^{(l,s)*}$ for the Lennard-Jones (12-6) potential. J. Chem. Phys. 57(3): 1100-1102 (1972).

16. P. J. Mohr, and B. N. Taylor, CODATA recommended values of the fundamental physical constants: 1998. J. Phys. Chem. Ref. Data. 28(6): 1713- (1999).

17. G. A. Olchowy, and J. V. Sengers, A simplified representation for the thermal conductivity of fluids in the critical region. Int. J. Thermophys. 10(2): 417-426 (1989).

18. R. Krauss, V. C. Weiss, T. A. Edison, J. V. Sengers, and K. Stephan, Transport properties of 1,1-difluoroethane (R152a). Int. J. Thermophys. 17(4): 731-757 (1996).

19. M. L. Huber, A. Laesecke, and R. A. Perkins, Transport properties of n-dodecane. Energy & Fuels. 18: 968-975 (2004).

20. J. V. Sengers, Transport properties of fluids near critical points. Int. J. Thermophys. 6(3): 203-32 (1985).

21. E. P. Sakonidou, H. R. van den Berg, C. A. ten Seldam, and J. V. Sengers, The thermal conductivity of an equimolar methane - ethane mixture in the critical region. J. Chem. Phys. 109(2): 717-736 (1998).

22. T. H. Chung, L. Ajlan, L. L. Lee, L. L., and K. E. Starling, Generalized multiparameter correlation for nonpolar and polar fluid transport properties. Ind. Eng. Chem. Res. 27: 671-679 (1988).

23. A. S. Cullick, and M. L. Mathis, Densities and viscosities of mixtures of carbon dioxide and n-decane from 310 to 403 K and 7 to 30 MPa. J. Chem. Eng. Data. 29: 393-396 (1984).

24. M. L. Huber, and J. F. Ely, Extension of an improved one-fluid conformal solution theory to real fluid mixtures with large size differences. Int. J. Thermophys. 11(1): 87-96 (1990).

25. A. Aucejo, M. Cruz Burguet, R. Muñoz, and J. L. Marques, Densities, viscosities, and refractive indices of some n-alkane binary liquid systems at 298.15 K. J. Chem. Eng. Data. 40: 141-147 (1995).

26. R. Heide, Viskosität von flüssigen hfkw-kältemitteln und deren gemischen. 225-241 (1996).

27. A. Laesecke, Viscosity measurements and model comparisons for the refrigerant blends R-410a and R-507a. ASHRAE Transactions: Symposia. 503-521 (2004).

28. A. Laesecke, R. F. Hafer, and D. J. Morris, Saturated-liquid viscosity of ten binary and ternary alternative refrigerant mixtures. Part I: Measurements. J. Chem. Eng. Data. 46: 433-435 (2001).

29. D. Ripple, and O. Matar, Viscosity of the saturated liquid phase of six halogenated compounds and three mixtures. J. Chem. Eng. Data. 38: 560-564 (1993).

30. C. Yokoyama, T. Nishino, and M. Takahashi, Viscosity of gaseous mixtures of HFC-125+propane from 298.15 to 423.15 K at pressures to 6.7 MPa Int. J. Thermophys. 27(3): 714-728 (2006).

31. K. Nagaoka, Y. Yamashita, and Y. Tanaka, Viscosity of binary gaseous mixtures of fluorocarbons. J. Chem. Eng. Japan. 19: 263-267 (1986).

32. J. Kestin, and S. T. Ro, The viscosity of carbon-monoxide mixtures with four gases in the temperature range 25-200 C. Supplement. Ber. Bunsenges Phys. Chem. 87: 600-602 (1983).

33. C. Yokoyama, T. Nishino, and M. Takahashi, Viscosity of gaseous mixtures of HFC-125(pentafluoroethane) + HFC-134a(1,1,2-tetrafluoroethane)(difluoromethane). Fluid Phase Equilibria. 174(1-2): 231-240 (2000).

34. M. Trautz, As given in Golubev, i.F., 1970, Viscosity of Gases and Gas Mixtures. A Handbook. 1953: Israel Program Sci. Transl. 245.

35. H. Iwasaki, and J. Kestin, The viscosity of helium-argon mixtures. Physica. 29: 1345-1372 (1963).

36. J. Kestin, and J. Yata, Viscosity and diffusion coefficient of six binary mixtures. J. Chem. Phys. 49(11): 4780-4791 (1968).

37. Y. Abe, J. Kestin, and H. E. Khalifa, The viscosity and diffusion coefficients of the mixtures of four light hydrocarbon gases. Physica. 93A: 155-170 (1978).

38. D. E. Diller, Measurements of the viscosity of compressed gaseous and liquid nitrogen + methane mixtures. Int. J. Thermophys. 3(3): 237-249 (1982).

39. N. Tripathi, Densities, viscosities, and refractive indices of mixtures of hexane with cyclohexane, decane, hexadecane, and squalane at 289.15 K. Int. J. Thermophys. 26(3): 693-703 (2005).

40. A. Fenghour, W. A. Wakeham, V. Vesovic, J. T. R. Watson, and E. Vogel, The viscosity of ammonia. J. Phys. Chem. Ref. Data. 24: 1649-1667 (1995).

41. E. W. Lemmon, and R. T. Jacobsen, Viscosity and thermal conductivity equations for nitrogen, oxygen, argon, and air. Int. J. Thermophys. 25: 21-69 (2004).

42. M. Takahashi, S. Takahashi, and H. Iwasaki, Viscosity of gaseous chlorotrifluoromethane (R 13) under pressure. J. Chem. Eng. Data. 30: 10-4 (1985).

43. R. C. Reid, Prausnitz, J.M., Poling, B.E., The Properties of Gases and Liquids. Fourth ed., McGraw-Hill, New York. (1987).

44. J. Millat, V. Vesovic, and W. A. Wakeham, The viscosity of nitrous oxide and tetrafluoromethane in the limit of zero density. Int. J. Thermophys. 12(2): 265-273 (1991).

45. M. Takahashi, N. Shibasaki-Kitakawa, C. Yokoyama, and S. Takahashi, Gas viscosity of difluoromethane from 298.15 to 423.15 K and up to 10 MPa. J. Chem. Eng. Data. 40: 900-902 (1995).

46. V. Vesovic, W. A. Wakeham, G. A. Olchowy, J. V. Sengers, and J. T. R. Watson, The transport properties of carbon dioxide. J. Phys. Chem. Ref. Data. 19: 763-808 (1990).

47. M. Takahashi, Yokoyama, C., Takahashi, S., Viscosities of gaseous 1,2,2-trichloro-1,1,2-trifluoroethane (R113), 1,2-dichloro-1,1,2,2-tetrafluoroethane (R114) and chloropentafluoroethane (R115). Kagaku Kogaku Ronbunshu. 11(2): 155-161 (1985).

48. D. C. Dowdell, and G. P. Matthews, Gas viscosities and intermolecular interactions of replacement refrigerants HCFC 123 (2,2-dichloro-1,1,1-trifluoroethane), HCFC 124 (2-chloro-1,1,1,2-tetrafluoroethane) and HFC 134a (1,1,1,2-tetrafluoroethane). J. Chem. Soc., Faraday Transactions 1. 89(19): 3545-3552 (1993).

49. M. L. Huber, A. Laesecke, and H. W. Xiang, Viscosity correlations for minor constituent fluids in natural gas: N-octane, n-nonane and n-decane. Fluid Phase Equilibria. 224: 263-270 (2004).

50. B. Le Neindre, and Y. Garrabos, Measurements of the thermal conductivity of HFC-125 in the temperature range from 300 to 515 K and at pressures up to 53 MPa. Int. J. Thermophys. 20(5): 375-399 (1999).

51. D. G. Friend, H. Ingham, and J. F. Ely, Thermophysical properties of ethane. J. Phys. Chem. Ref. Data. 20(2): 275-347 (1991).

52. H. Nabizadeh, and F. Mayinger, Viscosity of gaseous R123, R134a, and R142b. High Temperatures - High Pressures. 24: 221-230 (1992).

53. D. G. Friend, J. F. Ely, and H. Ingham, Tables for the thermophysical properties of methane. NBS Technical Note 1325. (1989).

54. E. Vogel, C. Kuchenmeister, and E. Bich, Viscosity correlation for isobutane over wide ranges of the fluid region. Int. J. Thermophys. 21(2): 343-356 (2000).

55. J. C. McCoubrey, and N. M. Singh, The vapor phase viscosities of the pentanes. J. Phys. Chem. 67: 517-8 (1963).

56. E. Vogel, C. Kuchenmeister, E. Bich, and A. Laesecke, Reference correlation of the viscosity of propane. J. Phys. Chem. Ref. Data. 27(5): 947-970 (1998).

57. Z. Shan, S. G. Penoncello, and R. T. Jacobsen, A generalized model for viscosity and thermal conductivity of trifluoromethane (R-23). ASHRAE Transactions. 106: 1-11 (2000).

58. M. Takahashi, S. Takahashi, and H. Iwasaki, Viscosity of gaseous chlorodifluoromethane (R-22). Kagaku Kogaku Ronbunshu. 9: 482-484 (1983).

59. C. Yokoyama, T. Nishino, and M. Takahashi, Viscosity of gaseous mixtures of HFC-134a (1,1,1,2-tetrafluoroethane) + HFC-32 (difluoromethane). 25(1): 71-88 (2004).

60. C. Yokoyama, T. Nishino, and M. Takahashi, Viscosity of gaseous mixtures of HFC-125+HFC-32 Int. J. Thermophys. 22(5): 1329-1347 (2001).

61. L. B. Bicher, and D. L. Katz, Viscosities of the methane propane system. Ind. Eng. Chem. 35(7): 754-761 (1943).

62. S. Chuang, P. S. Chappelear, and R. Kobayashi, Viscosity of methane, hydrogen, and four mixtures of methane and hydrogen from -100 C to 0 C at high pressures. J. Chem. Eng. Data. 21(4): 403-411 (1976).

63. J. G. Giddings, J. T. F. Kao, and R. Kobayashi, Development of a high-pressure capillary-tube viscometer and its application to methane, propane, and their mixtures in the gaseous and liquid regions. J. Chem. Phys. 45(2): 578-586 (1966).

64. I. M. Abdulagatov, and S. M. Rasulov, Viscosity of n-pentane, n-heptane and their mixtures. Ber. Bunsenges Phys. Chem. 100: 148-154 (1996).

65. P. Schley, M. Jaesche, C. Kuchenmeister, and E. Vogel, Viscosity measurements and predictions for natural gas. Int. J. Thermophys. 25(6): 1623-1652 (2004).

66. N. L. Carr, *Viscosities of natural gas components and mixtures*. 1953, Institute of Gas Technology Research Bulletin 23: Chicago, IL USA.

67. M. J. Assael, N. K. Dalaouti, and V. Vesovic, Viscosity of natural-gas mixtures: Measurements and prediction. Int. J. Thermophys. 22(1): 61-71 (2001).

68. L. I. Langelandsvik, S. Solvang, M. Rousselet, I. N. Metaxa, and M. J. Assael, Dynamic viscosity measurements of three natural gas mixtures-comparison against prediction models. Int. J. Thermophys. 28: 1120-1130 (2007).

69. J. Patek, J. Klomfar, L. Capla, and P. Buryan, Thermal conductivity of carbon dioxide - methane mixtures between 300 and 425 K and at pressures up to 12 MPa. Int. J. Thermophys. 26(3): 577-592 (2005).

70. J. Kestin, S. T. Ro, and Y. Nagasaka, The thermal conductivity of mixtures of methane with carbon dioxide. Ber. Bunsenges. Phys. Chem. 86: 945-948 (1982).

71. N. Imashi, and J. Kestin, Thermal conductivity of methane with carbon monoxide. Physica. 126A: 301-307 (1984).

72. W. J. S. Smith, L. D. Durbin, and R. Kobayashi, Thermal conductivity of light hydrocarbons and methane-propane mixtures at low pressures. Journal of Chemical Engineering Data. 5(3): 316-321 (1960).

73. H. M. Roder, and D. G. Friend, Thermal conductivity of methane-ethane mixtures at temperatures between 140 and 330 k and at pressures up to 70 mpa. Int. J. Thermophys. 6(6): 607-17 (1985).

74. O. B. Tsvetkov, Y. A. Laptev, and A. G. Asambaev, The thermal conductivity of binary mixtures of liquid r22 with r142b and r152a at low temperatures Int. J. Thermophys. 17(3): 597-606 (1996).

75. S. H. Kim, D. S. Kim, M. S. Kim, and S. T. Ro, The thermal-conductivity of R22,R142b, R152a, and their mixtures in the liquid-state Int. J. Thermophys. 14(4): 937-950 (1993).

76. R. A. Perkins, E. Schwarzberg, and X. Gao, Experimental thermal conductivity values for mixtures of R32, R125,R134a, and propane. 1999, NISTIR 5093.

77. S. T. Ro, D. S. Kim, and S. U. Jeong, Liquid thermal conductivity of binary mixtures of difluoromethane (R32) and pentafluoromethane (R125). Int. J. Thermophys. 18(4): 991-999 (1997).

78. Y. Tanaka, S. Matsuo, and S. Taya, Gaseous thermal conductivity of difluoromethane (HFC-32) pentafluoromethane (HFC-125) and their mixtures. Int. J. Thermophys. 16(1): 121-131 (1995).

79. X. Gao, Y. Nagasaka, and A. Nagashima, Thermal conductivity of binary refrigerant mixtures of HFC-32/125 and HFC-32/134a in the liquid phase Int. J. Thermophys. 20(5): 1403-1415 (1999).

80. D. S. Kim, M. H. Yang, M. S. Kim, and S. T. Ro. Thermal conductivities of pentafluoroethane (R125) and its mixtures with difluoromethane (R32) in the liquid phase. Fourth Asian Thermophysical Properties Conference. Tokyo, Japan (1995).

81. M. Yorizane, S. Yoshimura, H. Masuoka, and H. Yoshida, Thermal conductivities of binary gas mixtures at high pressures: N_2-O_2, N_2-Ar, CO_2-Ar and CO_2-CH_4. Ind. Eng. Chem. Fundam. 22: 458-463 (1983).

82. J. Kestin, Y. Nagasaka, and W. A. Wakeham, The thermal conductivity of mixtures of nitrogen with methane. Ber. Bunsenges. Phys. Chem. 86: 632-636 (1982).

83. S. T. Ro, J. Y. Kim, and D. S. Kim, Thermal conductivity of R32 and its mixtures with R134a. Int. J. Thermophys. 16(5): 1193-1201 (1995).

84. S. U. Jeong, M. S. Kim, and S. T. Ro, Liquid thermal conductivity of binary mixtures of pentafluoroethane (R125) and 1,1,1,2-tetrafluoromethane (R134a). Int. J. Thermophys. 20(1): 55-62 (1999).

85. S. U. Jeong, M. S. Kim, and S. T. Ro, Liquid thermal conductivity of ternary mixtures of difluoromethane (R32), pentafluoroethane (R125) and 1,1,1,3-tetrafluoromethane (R134a). Int. J. Thermophys. 21(2): 319-328 (2000).